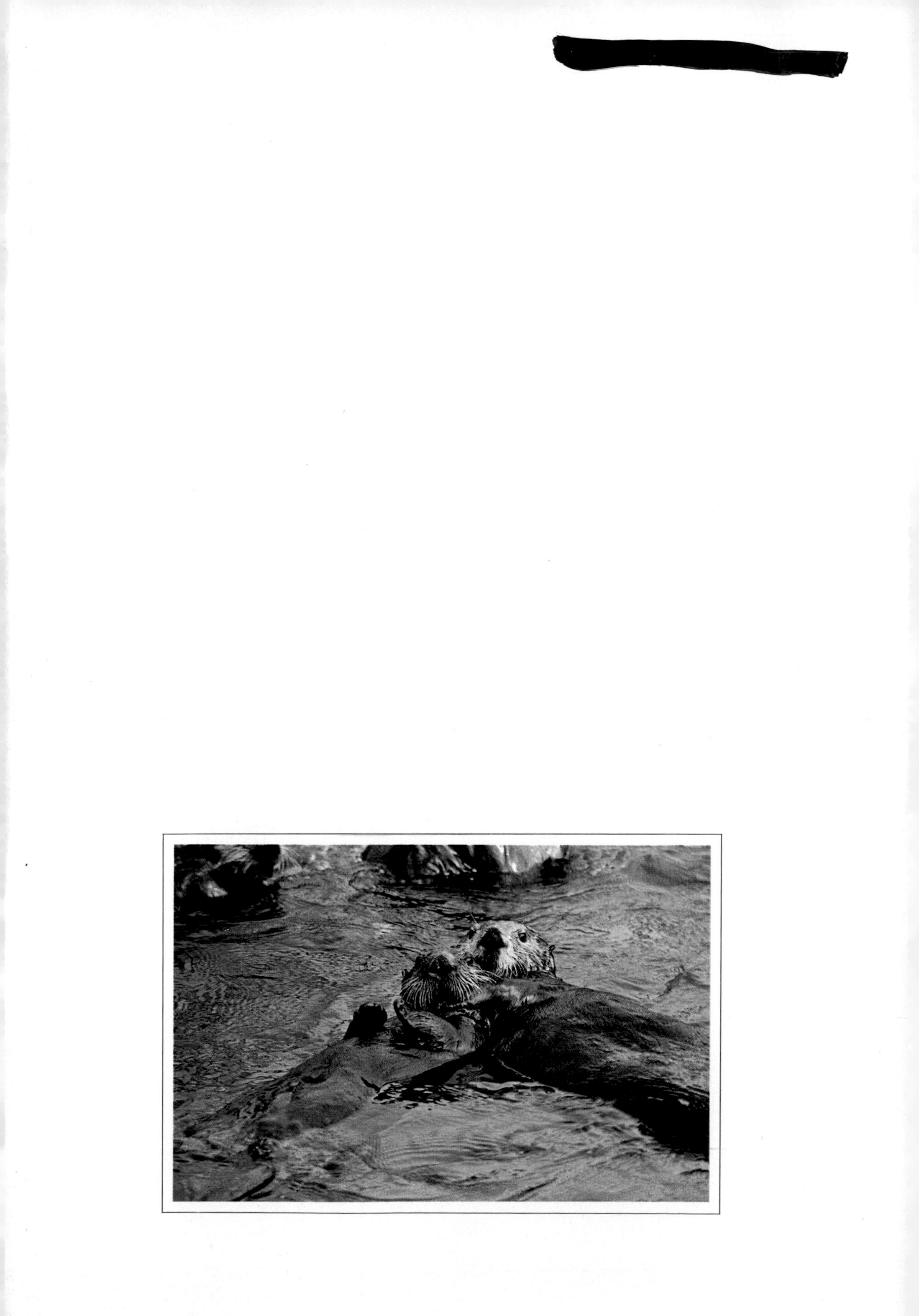

Vanishing Wildlife of North

America

By Thomas B. Allen
Foreword by Gilbert M. Grosvenor

Prepared by the Special Publications Division, National Geographic Society, Washington, D.C.

Vanishing Wildlife of North America

By Thomas B. Allen
National Geographic Staff

Published by
The National Geographic Society
Melvin M. Payne, *President*
Melville Bell Grosvenor, *Editor-in-Chief*
Gilbert M. Grosvenor, *Editor*
William Graves, *Consulting Editor*
Dr. Richard G. Van Gelder, *Consultant, Chairman and Curator, Department of Mammalogy, American Museum of Natural History*

Prepared by
The Special Publications Division
Robert L. Breeden, *Editor*
Donald J. Crump, *Associate Editor*
Philip B. Silcott, *Senior Editor*
Mary Ann Harrell, *Managing Editor*
Jan Nagel Clarkson, Marjorie W. Cline, Antoinette Eugene, Jennifer C. Urquhart, *Researchers*

Illustrations
David R. Bridge, *Picture Editor*
Josephine B. Bolt, David R. Bridge, Linda M. Bridge, Antoinette Eugene, William R. Gray, Barbara Grazzini, Wendy Van Duyne, Marilyn L. Wilbur, *Picture Legends*

Design and Art Direction
Joseph A. Taney, *Art Director*
Josephine B. Bolt, *Associate Art Director*
Ursula Perrin, *Design Assistant*
Paintings by Jay H. Matternes *and* Arthur Lidov

Production and Printing
Robert W. Messer, *Production Manager*
George V. White, *Assistant Production Manager*
Raja D. Murshed, Nancy W. Glaser, *Production Assistants*
John R. Metcalfe, *Engraving and Printing*
Mary L. Bernard, Jane H. Buxton, Marta Isabel Coons, Carol A. Enquist, Suzanne J. Jacobson, Penelope A. Loeffler, Joan Perry, Marilyn L. Wilbur, *Staff Assistants*
Virginia S. Thompson, Brit Aabakken Peterson, *Index*

Library of Congress CIP Data: page 207

Overleaf: An Everglade kite brings a snail snipped from its shell to her nestlings. Page 1: Sea otters rollick in Pacific shallows. Endpapers: Off California's coast, brown pelicans search for fish. Bookbinding: Whooping cranes, tule elk, and cutthroat trout represent threatened wildlife of sky, land, and water.

OVERLEAF: NOEL F. R. SNYDER; PAGE 1: N.G.S. PHOTOGRAPHER ROBERT F. SISSON; ENDPAPERS: TOM MYERS; BOOKBINDING SILHOUETTE DRAWINGS BY ROSALIE SEIDLER

TOM McHUGH, PHOTO RESEARCHERS, INC.

Gone from its namesake state of Michigan, a wolverine clambers up a tree in a Canadian forest.

Foreword

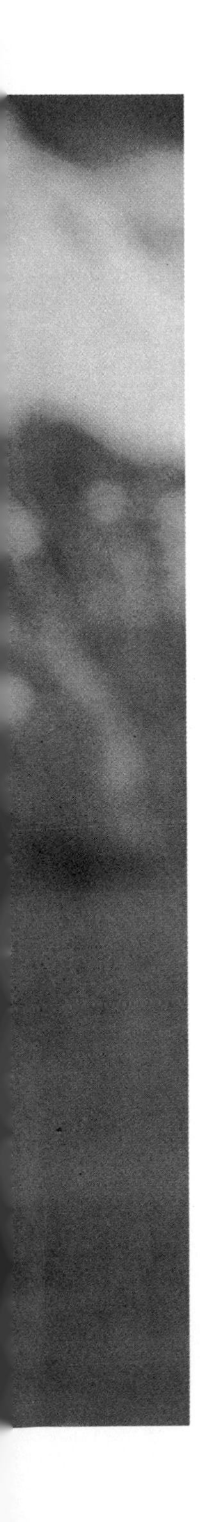

Did you ever wish that after your human life was over you could be reborn as a favorite animal? I remember playing that game as a child. A bald eagle appeared each year near my family's summer home at Baddeck, Nova Scotia. That's how I wanted to return—as an eagle—that particular eagle. How I wished that I could soar as a cloud-high sentinel above the boundless sea! Nothing could touch that eagle. He was a tough old bird with a firm hold on life.

I don't know what happened to my eagle, but I do know that eagles are losing their hold on life. Once their nests were abundant along the Great Lakes, at an estimated average of one for every ten miles of shoreline. Now bird watchers can count fewer than 20 active nests around Erie and Superior.

We may have acted—just in time—to save the eagle. In 1972 the Federal Government banned widespread use of DDT, implicated in the decline of this and other birds. Landowners protect eagles' nests on more than 3,000,000 acres in Florida. Biologists even prop up tall dead trees that serve as nest sites.

As this book goes to press, Congress has passed and the President has signed a new endangered species law. It requires all federal agencies to use whatever powers they have to protect our ecosystems. It requires certain key officials to disclose the reasons for their actions affecting imperiled wildlife. It gives citizens new power to make their voices heard.

But neither the eagle nor any other native animal can really be called safe. *Vanishing Wildlife of North America* is a title that states an indisputable fact. The very scope of our civilization menaces countless species—some as cherished as the eagle, some as easily overlooked as toads or turtles.

I learned how easy it is to overlook turtles on the day that I took my family to the Chincoteague National Wildlife Refuge in Virginia. We had gone there especially to show our son the famed wild ponies which roam the dunes of the refuge. What we did not see were the Atlantic loggerhead turtles that had been transplanted to its sands. Turtle eggs had been dug out of the sands of Cape Romain National Wildlife Refuge in South Carolina and taken to Chincoteague, where they hatched. If all goes well, in six to ten years hatchlings that crawled to the sea will return as mature adults. Man had driven the turtles away from their ancestral Virginia sands about twenty years ago. Now man is trying to re-establish the colony.

Perhaps when my son returns to Chincoteague he will see those turtles come home again. We owe our children a future with turtles and eagles and all the wild things that make the world so wondrous.

GILBERT M. GROSVENOR

THOMAS J. ABERCROMBIE, NATIONAL GEOGRAPHIC STAFF

Contents

An Alaska trapper displays pelts of ermine, lynx, marten, mink, red fox, river otter, wolverine—bounty and toll of a frontier wilderness.

1. The Cost of Conquest: a Continent Transformed

From a fearsome Eden through a century of extinction— to decades of attempted restoration

Two solemnly grinning Cheshire-cat faces stared down at me. I stared back long enough to see two more. No, wait. *Five*. I counted five bobcats draped on the shadowy limbs of the ancient live oaks that lined the trail. Farther along, I saw a two-week-old buffalo calf nuzzling its mother, two black bear cubs cuffing each other, three wolves loping along the edge of a wood, two elk bowing their heavy heads to graze, several deer dappled by sunlight and shade, and a tangle of alligators lazing in a marsh. A pure white egret glided through the dark canopy of the trail.

Then, around the bend, the illusion of a walk in the wild collapsed. I stood at a fence and watched two mountain lions—pumas, they're called here—pacing along a gully. Elsewhere in the "animal forest" at Charles Towne Landing, a historical park in Charleston, the illusion is maintained by ingeniously camouflaged barriers. But in reconstructing the past, the South Carolina Tricentennial Commission had to recognize the present. We are no longer one with the puma, which Indians here revered as a creature touched with divinity.

I had come to this pocket of the past to see animals the

Near her flock, a wild turkey seeks food in a Wyoming ponderosa pine forest—not part of the original range, but an example of management success in saving a bird once abundant across the land.

RON KLATASKE

JOSEF MUENCH

Primeval and pristine, North America once teemed with wildlife; fragments of landscape recall that inheritance today.

Around the wind-molded dunes in Death Valley National Monument flourishes an astonishing variety of plants and animals, each in a delicate balance with its forbidding habitat. Desert ecosystems like this, though comprised of tough and well-adapted individuals, remain so fragile that damage to one of their interdependent species can swiftly change or destroy the whole.

More resilient, but also more easily exploited, forests took the brunt of early and heedless use.

In autumn foliage, this virgin stand of trembling aspen on the eastern scarp of California's Sierra Nevada has escaped ax, saw, and more ponderous machinery. Cut only by beavers, felled trees dam a tranquil pond.

Beyond such pockets of wilderness, the continent keeps few areas comparatively unchanged and unharmed: in Mexico, the Baja peninsula; in Canada or Alaska, backwoods and tundra.

settlers saw when they landed in 1670 to found the first English colony in South Carolina. From Charles Towne Landing I could look westward to a conquered continent and backward to the time when that conquest began. Around me in the animal forest were the reminders and the remnants of a human triumph that was a tragedy for wildlife.

South Carolina was a fearsome Eden. The colonists believed, as one noted in 1709, they lived amid "endless Numbers of Panthers, Tygers, Wolves, and other Beasts of Prey" that filled the night with "the dismall'st and most hideous Noise." But there was meat aplenty: wild pigeons in flocks so dense they blotted out the sun; deer, elk, and buffalo in the backcountry.

The elk and buffalo and deer I saw were actors brought on stage to portray the past. Neither the elk nor the buffalo roam South Carolina today; they were wiped out. Deer live in the state today largely as transplants, restocked for hunters. The last native wolf in South Carolina disappeared sometime about the middle of the 19th century. The pumas I watched were imports from South America and the West.

As for the others I saw in the animal forest, each can claim a living place in South Carolina today. But each, too, is a symbol of the tragedy that still unfolds. The bear fell from lordly beast of virgin forest to cartoon character and national park beggar. The egret, for its plumes, and the alligator, for its hide, nearly died out as species in an orgy of shooting. The bobcat, trapped for its pelt and classified as a bounty animal in at least 18 states, lives in uneasy coexistence with civilization.

Driving along a throughway out of Charleston, I saw this for myself. A bobcat kitten crouched by the side of the road, alone. About three miles down the road I passed by what I sadly assumed was the body of the kitten's mother.

To see two bison or a couple of elk and then try to conjure the herds of the past is to see a single tree and try to imagine a forest, to see a single petal and try to imagine an incredibly vast garden rolling beyond the horizon. That past time, which we cannot summon, was still the future which the Charles Towne settlers could not perceive.

They had taken their first steps onto a continent that teemed with wildlife. The New World sustained tens of millions of bison, perhaps the biggest array of such a species ever to tread the earth. Elk, which had flourished over most of the continent, were already in retreat from the forests being cleared in the colonies of New Netherland and New England. Deer were easy to take and the supply seemed inexhaustible. By 1699, when the Carolina colony shipped 64,888 deerskins to England, the conquest had begun.

To the north, an empire was being built on beaver skins. In 1670, the very year that his namesake town was founded, Charles II granted a royal charter that established the empire, the Hudson's Bay Company, for a century and a half the only government over an area bigger than Europe.

Seeking the beaver, *Castor canadensis*, company agents

JOSEF MUENCH

Gnarled in years of wind, Monterey cypresses thrive in thin soils by California's Carmel Bay. Botanists consider them an endangered species—their coastal habitat invites home-builders.

*To protect a fine natural stand of these trees as well as imperiled animals, the legislature in 1933 created Point Lobos State Reserve just three miles south. It shelters two species (Steller, and California) of sea lions—*los lobos marinos, *or wolves of the sea, to Spanish explorers.*

Overlead: A lone buffalo, silhouetted where mountains meet grassland on the Flathead Indian Reservation in Montana, evokes an age when hunted and hunter alike roamed numerous and free.

LOWELL GEORGIA

staked out boundaries for a dominion that would make a national symbol of an animal the fur trade nearly exterminated. Beaver skins became the coin of its realm; twelve bought a rifle, six a blanket. And from the naming of Beaver Brook in Massachusetts in 1632 to the founding of Beaver, Washington, in 1891, the massacre of an animal preceded the peopling of a continent. The fur trade, which also took a host of other species, set the pattern of outpost become city (Victoria, Edmonton, Detroit, Albany, Pittsburgh, to name a few).

Another tradition grew from fear and hatred of predators. The Massachusetts Bay Company put a penny-a-pelt bounty on the wolf in 1630. William Penn bountied the wolf in 1683. South Carolina, in its 1695 "Act for Destroying Beasts of Prey," ordered every Indian bowman to hand over each year either two bobcat skins or the pelt of a wolf, of a panther, or a bear. Laggards were to be "severely" whipped. By the Revolutionary War panthers were almost gone from southern New England; the wolf was nearly extinct there and in eastern Canada.

If wolves and mountain lions were foes to be slain, birds were friends to be slain—but for food. The state that would emerge from the Charles Towne colony would give to science more descriptions of new species than any other state. In 1664 a visitor reported that "great flocks of Parrakeeto's" filled the skies. The country abounded with "Turkeys, Quails, Curlues, Plovers, Teile, Herons; and as the *Indians* say, in Winter, with Swans, Geese, Cranes, Duck and Mallard, and innumerable of other water-Fowls, whose names we know not...."

But in less than 200 years North Americans would wipe out at least nine of their continent's bird species—a record of extinctions unmatched in any other large area on earth. We still do not know the exact number because some species, such as the ivory-billed woodpecker and the Eskimo curlew, exist at least in hopes sustained by reported sightings.

The extinctions began with the great auk, a big, flightless bird that bred on rocky specks in the North Atlantic. When man saw the auk, it was death at first sight. Jacques Cartier landed on one of their breeding rocks, Funk Island off Newfoundland, in 1534, and killed enough in half an hour to fill two longboats. The auks survived numerous raids by ships' crews hungering for meat, but in the era of settlement, the birds became the raw materials of an industry. They were boiled for their oil. When a Norwegian naturalist landed there in 1841, all he found were debris and bones.

About the time the auk was disappearing, the spectacled cormorant was being exterminated on a desolate island in the Bering Sea. This beautiful bird—a greenish-blue crest and long yellow feathers crowned its blue-and-green body plumage—was seen by naturalist Georg Wilhelm Steller in 1741. Within a century of its discovery, it was eaten into extinction by a

Almost destroyed by trappers in centuries past, the beaver has regained former ranges by the work of wildlife managers. Its dams, like this one in Wyoming, aid conservation of soil, water, and other species.

JEN AND DES BARTLETT

new branch of the fur trade—hunters of seals and sea otters.

The slaughter of the animals of the sea stained both Atlantic and Pacific waters in the 19th century, but the gravest threat of extinction flew with the immense flocks that coursed the skies of the continent. The passenger pigeon may have been the most plentiful bird man has ever known. John James Audubon once calculated that a single flock contained 1,115,136,000. They were killed by the tens of thousands, and not even Audubon realized that someday the last, last one would die. "The passenger pigeon needs no protection," a select committee of the Ohio state senate reported in 1857.

The Carolina parakeet also seemed phoenix-like, ever rising, year after year, from fields made charnel houses by the guns of farmers and market hunters. Then, state by state, the birds of market disappeared. And then there were none. The Cincinnati zoo boasted the last bird of both species—the final pigeon in 1914, the final parakeet in 1918.

Other birds were dying for fashion. In 1886, an ornithologist shifted his field studies to the streets of New York City. During two strolls he observed 700 ladies' hats, 542 of them adorned with birds of 40 different species. In the early 1900's, the common egret's back feathers—the filmy aigrettes of its nuptial plumage—sold for $32 an ounce, nearly twice the price of gold.

From Cape Cod, where plume hunters killed 40,000 terns a year, to Florida, where one feather merchant wholesaled 130,000 skins in a single year, birds were dying not for food but for "man's greed and woman's vanity," as one scientist angrily declared. Clamor against the carnage evolved into action. The American Ornithologists' Union was born and pointedly named its publication *The Auk*. Audubon Societies formed a national federation. Critics of the plume trade managed to get some protective laws passed, but not enforced.

The magnificent birds of the Everglades—the egrets, the herons, the roseate spoonbill (worth $5 a skin)—were nearing extinction when the Audubon Societies hired four wardens to guard a dwindling number of birds from a growing army of hunters. It seemed a quixotic gesture—four unlikely knights in swamps or islands pitted against the world of high fashion.

But warden Guy Bradley would doom an industry worth millions. On July 8, 1905, he spotted a suspicious schooner near Oyster Key, off the tip of the Florida mainland. He set out in his skiff and reached the ship just as two hunters were loading four dead egrets aboard. When he tried to arrest them, the skipper shot him and set sail, leaving the warden dead in his drifting boat. Bradley's killer, claiming self-defense, went free.

Beyond the backwoods courtroom, the outraged verdict was murder. President Theodore Roosevelt added his voice to the outcry. Two years earlier he had authorized on four scant acres the first federal wildlife refuge, at Pelican Island, Florida. The killing of Bradley mocked such feeble measures. Horrified citizens demanded new and stronger laws. It took five years for the legislators in New York, center of the millinery business, to pass a law that effectively ended *(Continued on page 24)*

Warm, safe, and cozy inside a lodge near Orem, Utah, a beaver nurses four of her two-week-old kits, apparently ignoring the lights installed under the roof or the photographers' presence in an adjoining plywood box screened by a black curtain.

Able to swim from birth, kits younger than these find it impossible to dive; their thick fur traps too much air. Once their body weight increases to a few pounds, they will leave the lodge by an underwater passage—as does an adult in the picture above.

Beginning to explore outside, they nibble leaves and twigs, the staples of the vegetarian beaver. Kits born in spring can work for the colony by autumn; they help to hoard food and repair lodge and dam before winter.

Throughout life the beaver's incisor teeth continue to grow. Constantly honed by gnawing, they enable one animal to bring down a three-inch tree in 15 minutes. Saplings or branches cut to manageable length and roughly piled make a framework for the lodge, caulked with mud and floored with bark, or for dams sturdy enough to hold back acres of water.

To keep their coats healthy, beavers have a "combing claw" on each hind foot and oil glands near the tail. The yellow castoreum from two musk glands—prized as a fixative for perfumes and as a cure-all in the past—and the beautiful pelts brought the animal close to extinction in the Old World and the New.

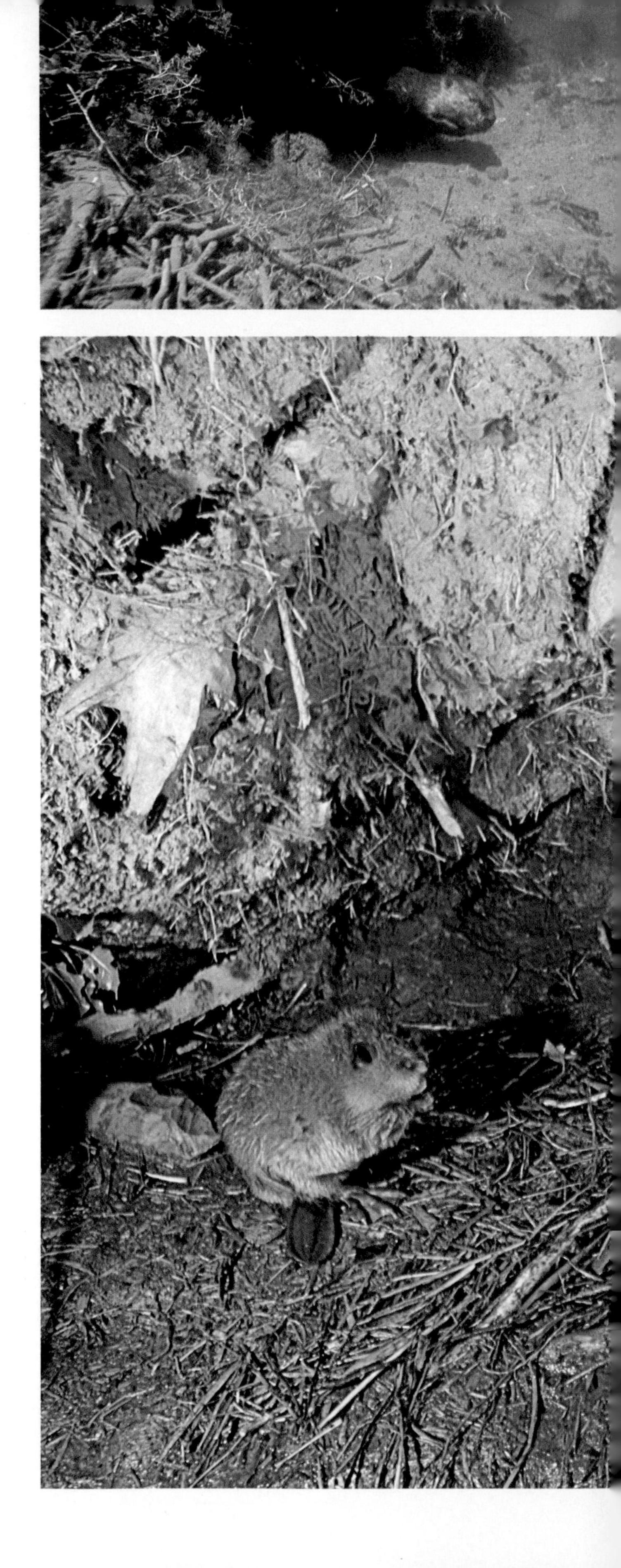

JEN AND DES BARTLETT

Now only an emblem of extinction, the great auk bred on islands of the North Atlantic—safe from all predators before the sailors came. At right, a breeding pair exchange greetings above their single egg. On the cliffs behind them, other sea birds take roosting and nesting sites at different levels according to species. The auk shared its rookeries with a smaller cousin, the common murre (above), which survived through its ability to fly. When diving for fish, the 30-inch auks kept their little wings outstretched as stabilizers.

These flightless birds, clumsy ashore but swift and powerful swimmers, migrated as far south as Florida in winter, as far east as the British Isles. In an age of discovery and settlement, men found auk rookeries a source of provision and profit. Batted along with oars, herded waddling into pens and bludgeoned, the unresisting auks furnished food, featherbeds, and even oil as fuel for their own scalding. As they became rare, collectors scurried to claim and stuff the few remaining specimens and gather eggs for museums. Since about 1844 no one has seen the birds alive.

Discovery and destruction: Harpoon in hand, a Russian sailor stalks a sea cow—food for the starving crew of explorer Vitus Bering's St. Peter, *wrecked near Alaskan waters in 1742. Ashore, men with stout rope wait to haul in the struggling giant. Delicious meat proved a margin of survival for building a new ship. The expedition's naturalist, Georg Wilhelm Steller, noted that these sea mammals lived in herds; they fed together on kelp. When the men*

PAINTINGS BY JAY H. MATTERNES

hooked one, all the others tried to save it, circling the victim or jostling the boat; for two days a male swam in near its mate lying dead on the beach. Sealers and traders hunted Steller's sea cow into extinction within 27 years.

Below, a 30-foot adult—painted from Steller's notes—and a diver give scale to the surviving members of the order Sirenia: the 12-foot dugong (right) of tropical Asia and the 15-foot manatee of the southeastern United States.

the feather trade. Revulsion against the exploitation of animals inspired an abundance of statutes, but none could bring back the abundance of wildlife the continent had known.

By 1900, the pronghorn, the so-called American antelope, had vanished from 80 percent of its range; in most of the country east of the Rockies there were no more deer to hunt. The great herds of elk had vanished from the East, leaving only memories in place names—Elk Creek in New York, Elk Neck in Maryland. Shrunken herds found a few havens, such as Yellowstone, established in 1872 as the first national park, and a private refuge for the tule elk in California. But nowhere could they find permanent sanctuary from market hunters and unregulated sportsmen.

Of all the species straggling toward extinction in 1900 none summoned more sorrow and shame than the bison, the American buffalo, the symbol of the unconquered West. More than the buffalo fell before the advancing frontier, for it was but part of a vast ecosystem linking grass, buffalo, and Indian. Those first buffalo hunters got from their prey not only meat but also sinews for bowstrings, hides for tepees and clothing. The buffalo kept the Plains Indians alive.

Even though 19th-century Americans did not know the intricacies of an ecosystem, some had enough perception to see that the eradication of the buffalo would hasten the eradication of the Indian and the opening of the prairie for cattle. When both Houses of Congress passed a bill in 1874 to save the buffalo, President Grant declined to sign it, apparently because his Indian fighters opposed it.

As the railroad and the rifle moved westward, the buffalo fell before them—for food, for war strategy, for sport, for bravado. Passengers lined the windows of trains to shoot at the animals. It was said you could walk along 100 miles of railroad right-of-way and never step off a buffalo carcass.

Some 60,000,000 buffalo roamed much of the continent before the white man arrived. By 1800, hardly any existed east of the Mississippi River, but an estimated 40,000,000 still survived in the West as late as 1830. For the next four decades they endured remorseless butchery. Their last days began in the 1870's, but even then they seemed to march without end. Near Fort Hays, Kansas, in September of 1871, a troop of the Sixth Cavalry came upon a herd first believed to number in the hundreds... then the hundreds of thousands. "For six days," the young commander reported, "we continued our way through this enormous herd, during the last three of which it was in constant motion across our path." He found it "impossible to approximate the millions."

Fifteen years after 2nd Lt. George S. Anderson saw that herd, Dr. William T. Hornaday of the National Museum toured the West in search of buffalo. He recorded 541, many of them at Yellowstone. In 1891, Anderson, by now a captain, became acting superintendent of the park. The millions, the hundreds of thousands, the thousands—now they were the 300.

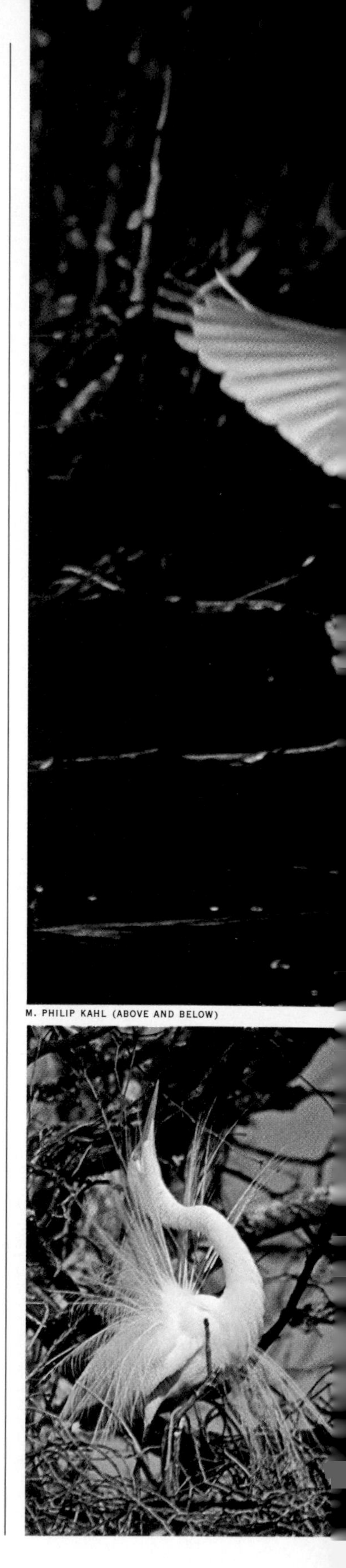

M. PHILIP KAHL (ABOVE AND BELOW)

JEN AND DES BARTLETT, BRUCE COLEMAN INC.

Alighting on a Florida pond, a common egret extends its wings to full 55-inch span. Aigrettes, filmy plumes of the mating season, adorn all adults for courtship (far left). In the early 1900's these feathers cost almost twice their weight in gold; plume hunters killing birds for millinery nearly obliterated the species, leaving eggs to rot and chicks to starve in the rookeries. Protected today by law, two adults squabble over the head of a younger bird.

"The buffalo is gone," Francis Parkman, historian of the West, wrote in 1892, "... wolves ... have succumbed to arsenic and hushed their savage music.... The mountain lion shrinks from the face of man, and even grim 'Old Ephraim,' the grizzly bear, seeks the seclusion of his dens and caverns."

Not gone, somehow not gone. Yellowstone, almost by chance, had become an ark for wildlife. In the deluge of civilization surging across the continent, the ark preserved the nation's last wild buffalo—down to 39 in the park by 1900. And within Yellowstone, effectively closed to hunters in 1894, other stocks of animals found refuge from the guns. The pronghorn seemed doomed, but the park's small herd would become a nucleus for later restorations. At a time when Merriam's elk in the Southwest and Audubon's bighorn of the Dakota badlands became extinct, the elk and bighorns of Yellowstone lived on. Wolves, mountain lions, bobcats, coyotes, and bears, all of them targets for extinction beyond the park, clung to life inside it.

Yellowstone was the first of a fleet of arks launched just in time to carry our wildlife through what biologists have called a century of extinction. The 20th century dawned on an era of conservation that saw the creation of national park and refuge systems, the emergence of strong animal-protection societies, and the passage of laws to preserve the animals that survived the conquest of the continent.

Many species that barely survived the 19th century would flourish under the care of game managers, a new breed of conservationists primarily employed to provide quarry for hunters and fishermen. White-tailed deer, near extinction in the late 1800's, now thrive in every state east of the Rockies. Probably they are more abundant now than when the settlers arrived to found Charles Towne. About 120,000 of them—and 1,200 other big-game animals—are killed each year on the highways.

The beaver, reprieved when the silk topper supplanted the fashionable hat made of felted beaver fur, was nearly gone at the turn of the century. Managers have brought it back, restocking it where it had vanished. Beavers have been parachuted into Montana wilderness, backpacked into Maine, taken in gunnysacks to places that needed their dams. The beaver population in Alabama alone has soared from 500 to 200,000 in 20 years.

The birds for which Guy Bradley gave his life still grace the Everglades. But both the birds and the Everglades are in danger. So are hundreds of species of wildlife—with the lands, waters, and the very air that sustain them. Conservationists who once worried about other species now worry about their own.

Endangered species have begun signaling a new kind of peril: environmental pollution by insecticides. When I interviewed Dr. W. Earl Godfrey, curator of birds at Canada's National Museum of Natural Sciences, he recalled one of the earliest danger signs. On a spring day in 1949, the late G. Harper Hall,

CHARLES HARBUTT, MAGNUM

common egret

great blue heron

One-day-old egret, crowned with fluff, crouches by eggs yet unhatched. Alert young herons, nestlings still at about three weeks, stiffen into immobility before a stranger—a protective response for concealment.

RON KLATASKE (BELOW AND LOWER LEFT)

great blue heron

an amateur ornithologist, was watching peregrine falcons that had nested on a tall building in Montreal since 1936. "What he saw was to him odd behavior," Dr. Godfrey said. "They were even eating their own eggs." Dr. Godfrey picked up one of his reports on endangered species from his desk and began reading it to me: ". . . these birds were, in reality, sick. Indeed, . . . Hall presented one of the earliest records of certain aberrant behavior which we now know to be symptomatic of birds carrying high residues of the chlorinated hydrocarbons such as DDT."

The peregrine falcon, almost obliterated over half the continent in 20 years, may be the passenger pigeon of our time. But other animals face extinction today because of a peril far older than DDT: immediate conflict with man.

Snow still cloaked most of Yellowstone National Park when I visited there on a day of awakening spring. I saw a few deer browsing in a snow-patched meadow and heard the faint song of unseen birds. The stillness reminded me of the hush that falls upon a theater in the moments before the curtain rises. Soon the wildlife of Yellowstone would come on stage. And soon would come the audience—some 2,000,000 people a year, drawn to the drama of a living wilderness.

"We want a visitor to enter the park, see the wildlife, and leave—without having done anything to change what is there," Glen F. Cole, the supervisory research biologist, told me. "By jumping out of a car and charging after an animal, a visitor keeps that animal from doing what it wants to do, what it is supposed to do." He gazed out the window of his office at the strangely empty park. "Existing fauna has been here for 30,000 years. They got along for quite a time before there were biologists and bureaucrats around."

Yellowstone is as much a stage of the present as Charles Towne Landing is a stage of the past. Many animals appear in both places. In the little animal forest at Charles Towne they portray memories; in the 2,221,760-acre national park they *live*. Yet even here two species are in conflict with man: the grizzly bear—and, I was startled to learn, the buffalo.

Each conflict involves a straightforward question: How best can visitors and grizzlies be protected from each other? Should the buffalo in Yellowstone be vaccinated to prevent them from spreading disease to cattle herds outside the park? But other questions haunt many of our vanishing species. What is the worth of animals? Is the death of a species—or a unique population of a species—ever justified?

On a single night in 1967 two young women were killed by grizzly bears in Glacier National Park, the first such fatalities since the park opened in 1910. Grizzlies in Glacier, as in Yellowstone, long had feasted on the park's garbage. In the shocked reaction that followed the Glacier deaths, the parks began closing their dumps and cutting off the bears from "unnatural" foods. Then, in 1972, Yellowstone recorded its first known killing of a person by a bear in 57 years. Glen Cole reported it thus: "A man returned to an unauthorized campsite in the dark and encountered a grizzly that was *(Continued on page 33)*

Extinct in 1918: The Carolina parakeet appears above with agents of its fate—enemies and admirers.

In misleading abundance these birds had ranged the southeastern and central United States, their primary habitat the cypress swamps and wooded river-bottom lands of the deep south. But their range contracted westward, diminishing in proportion to the increase of the human population.

Seed-eaters that had thrived on cockleburs, thistles, pine and cypress nuts, the parakeets soon turned to fruit—in orchards on newly cleared land. Destroying unripened oranges, apples, and peaches, they were destroyed in turn by outraged farmers.

Men protecting their groves would kill entire flocks when the unhurt birds returned to hover screaming over the dead and wounded, easy marks for the guns.

PAINTING BY ARTHUR LIDOV

In addition, brilliant plumage made the birds popular decorations for ladies' hats in the 1880's and '90's. Vast but unreckoned numbers died for millinery.

Gregarious habits left them vulnerable. In the wild, an entire colony would nest in a hollow tree. A burlap bag over the entrance hole and a sudden shout could capture them all alive.

Easily tamed in two days, intelligent and docile, these small parrots found eager buyers throughout North America and Europe. Caged and kept as pets, they usually died without issue.

Only belatedly did their status arouse concern as reports from the wild grew sporadic. "To this charming little . . . bird," wrote naturalist William T. Hornaday in 1913, "we are in the very act of bidding everlasting farewell." By nightfall on February 21, 1918, the last captive had died.

Buffalo in "innumerable hordes" cross a river bottom in search of water and food—so American artist William Jacob Hays explained for a London catalogue. In 1860 he had traveled up the Missouri River, and he warned that canvas could not "adequately convey" the herds. Once ranging in tens of millions, bison had furnished the Plains Indians with food and hides; in the Indian wars, generals considered extermination of the buffalo a strategic necessity. Millions fell for the hide trade, for their tongues (a delicacy), for sport. By 1900 only 39 wild bison survived in the United States—all in Yellowstone National Park. The herd there today numbers 800; at right, two rush through deep snow. Including animals privately owned, 30,000 exist in North America now.

WILLIAM J. HAYS, "HERD ON THE MOVE," THE THOMAS GILCREASE INSTITUTE, TULSA, OKLAHOMA; WILLIAM ALBERT ALLARD

374

eating groceries and garbage that had been left on the ground." Such careful wording typifies the accounts of man *v.* grizzly in Yellowstone, whose officials are well aware of the controversy that swirls around their bear-management program.

Dr. John J. Craighead and Dr. Frank C. Craighead, Jr., on the basis of an intensive study of Yellowstone bears begun in 1959, advised in 1967 that open-pit garbage dumps be gradually phased out, "enabling the grizzlies to develop new feeding habits as well as altered social behavior and movement patterns." Glen Cole, however, believed that bears rapidly cut off from human food sources would more quickly learn to rely on natural food. But those bears that did not learn the new rules would have to be transplanted to backcountry areas or, if they became incorrigible foragers in campsites, be disposed of.

The prospect of killing bears inflames the controversy. Conservation groups, such as the Fund for Animals, protest such management. And critics ask why the grizzly is hunted as a game animal and why it was removed from the official federal List of Endangered Native Fish and Wildlife. Perhaps as many as 100,000 grizzlies prowled the densest wildernesses of the West as late as 1850. Now fewer than 1,000 are believed to exist south of the Canadian border.

From 1970 through 1972, Cole reports, 13 park grizzlies were deliberately killed and 13 others died by mishaps while being transferred; 4 were accidentally killed by vehicles in the park. In the same period, Montana game officials killed 16 bears outside the park to "control" them, and 45 bears were killed—as legal game or "to protect livestock or property"—in Montana and in Idaho and Wyoming, the other states that border Yellowstone. The known total, then, from varied causes, is 91. But all 91 deaths have often been blamed "on Yellowstone."

Glen Cole winced when the dilemma of bear management came up in our conversation. He sees the grizzlies as inhabitants of what is becoming once more a natural ecosystem, where animals must fend for themselves. Wolves and mountain lions used to be "controlled," as was the resident herd of elk. "Now the wolf may be coming back," he told me. "They're here, but we don't think there are enough yet for them to have a social system. They need that to become a real resident population."

A limited food supply, harsh winters, disease, and predation are the principal natural checks on the park's herds of deer, elk, and buffalo. Dr. Mary M. Meagher of the park staff, a research biologist particularly concerned with the buffalo, looks upon the Yellowstone herd as a priceless American heritage and natural regulation of it as essential to preserving that heritage.

About 30,000 buffalo live on public and private lands elsewhere in the United States and Canada. In some places, surplus

With a rattle on a pole, a veterinarian moves Yellowstone buffalo at a holding pen before taking blood samples during a controlled reduction of the herd. Park officials discontinued such measures in 1966; current research leads them to think artificial controls unnecessary.

Bull elk at Yellowstone spar with sharp-tined antlers in combat for dominance after the rut. Winter forage may prove inadequate for elk and competing species: mule deer, pronghorns, bighorn sheep. At right, biologists round up elk by helicopter for transfer to new ranges; a stately bull takes position behind the cows and calves. Early in 1968 the park gave up this practice; it now relies on weather, food supply, migration, and hunting outside the park to limit the size of the elk herds.

GUS WOLFE (ABOVE); WILLIAM ALBERT ALLARD

animals (more than what managers call the "carrying capacity" of the habitat) are regularly sold for meat or offered as targets for hunters. Yellowstone's 800 buffalo are unique, as Mary Meagher emphasizes. "Only here in the entire U. S.," she says, "have wild free-ranging bison survived since primitive times. They have esthetic and scientific values which cannot be duplicated elsewhere in the country, and which cannot be replaced if the park bison are destroyed." This is what would happen, she believes, if her buffalo lose the battle over brucellosis.

Brucellosis, also known as Bang's disease, is dreaded by cattlemen—ranchers as well as dairymen. It causes cows to abort and lays human victims low with symptoms that resemble influenza or malaria. In human beings, the disease is rarely fatal; it responds to antibiotics. But, because of its disastrous effect on the livestock industry, the U. S. Department of Agriculture has launched a program to eradicate brucellosis throughout the nation as rapidly as possible. The campaign involves the killing of carriers and the inoculation of heifers with a vaccine that may give lifetime protection.

When local ranchers and the Department of Agriculture asked park authorities to round up the buffalo and kill any carriers, the herd's keepers refused. Mary Meagher knows that brucellosis is probably present in as many as 60 percent of her buffalo. But she believes that any attempt to eradicate brucellosis would probably mean destruction of the herd as a free-ranging wild population. She estimates that from identifying and then killing carriers, and shooting buffalo too inaccessible for inspection, only 40 animals would survive.

In Bozeman, Montana, about 50 rugged miles from the northern boundary of the park, I talked with dairymen who spoke passionately of the need to wipe out brucellosis in the Yellowstone herd. None, however, could tell me of a local case involving a park buffalo. Yellowstone authorities say they would shoot any strays that could get near cattle ranges. In fact, there is no known case of a wild buffalo infecting a domestic cow.

The drama at Yellowstone had turned out to be backstage drama: man against animal, Department of Agriculture against Department of Interior, even biologist against biologist. And fear of animals mingled with love of animals.

Looking back, beyond Yellowstone, beyond the conquest, back to those days of fear and hatred at Charles Towne, I realize that the proper study of wildlife is man. What we fear, what we hate, and what we admire in animals determine their fate. Consider *el lagarto,* the lizard that terrified Spanish explorers, "this amphibious Monster" to an English traveler. I had seen alligators in the animal forest. I would see them again, living free but as embattled as the buffalo. I would see some men killing them, while others pleaded that the alligator be saved.

Grizzly triplets, age about 16 months, scamper uphill in Yellowstone followed by their mother. Despite her look of benign clumsiness, she will defend a threatened cub with ferocity—and with incredible speed.

MAURICE G. HORNOCKER

2. “There Is No Ark This Time”: Dilemmas of the Lists

In clashes of law and policy and expert opinion, salvation for wildlife lacks certainty

Shortly after dawn, Old Bread rose from the dark waters of the bayou, summoned by the *cluck-cluck* sound that a man makes when he tries to call up an alligator. The oil workers who had named Old Bread made that sound on the days they tossed him the crusts of their sandwiches. But on this day, when his long snout appeared, a man on the shore leaned down and put a bullet through Old Bread's skull. His name would give him prominence among the first alligators legally killed by hunters in Louisiana in almost a decade, the first to die in a 13-day hunting season that began at his last dawn.

Old Bread's hide and the hides of 1,346 other alligators would be stacked in a shed at a state refuge, headquarters for the hunt and the place where I began learning about the plight of North America's vanishing wildlife. Here, where hunters confronted prey, I found another confrontation over an animal which the State of Louisiana called a renewable resource and the Federal Government called an endangered species. Here, in a state whose license plates proclaimed a "sportsman's paradise," I got my first look at a new kind of Eden, where man no longer lives with animals; he manages them.

Visiting observer in Louisiana, a South Carolina warden examines an alligator during a 1972 hunting season. The U. S. Government listed the gator as endangered—but could not control state game policy.

DAVID R. BRIDGE, NATIONAL GEOGRAPHIC STAFF

This hunt for the alligator in the marshes of southwestern Louisiana epitomized the entire continent's wildlife crisis. Here was an animal treated as a crop to be planted, thinned out, harvested. Here was the paradox of saving a species by "cropping" it. Here was the fear that a species snatched from the brink was being forced toward it again. Here were biologists trying to rescue animals from the snares of civilization. Here was a conflict between agencies responsible for animal welfare. And here was a predator that had endured for eons now facing the most successful predator ever to evolve: man.

Slaughter of the alligator rivaled the bison's. At least ten million were killed for their hides between the early 1800's and 1940. Around 1956, when the annual harvest sharply declined —as did the size of individuals—the nine alligator states were becoming concerned. In Louisiana, an apparent increase eased the worry—until a closer look showed that the floodwaters of Hurricane Audrey in 1957 had carried thousands of gators out of refuges and into areas where they had no protection.

One such state sanctuary was the Rockefeller Wildlife Refuge, about 60 miles east of the Texas border. Ted Joanen, research leader for the state refuges, has helped to restore the alligators there. He admires the alligator, and he wants its safety assured. But, like the majority of his management colleagues in Louisiana, he does not believe it is endangered.

"We know gators here," he said when he was briefing me on the hunt. "Ninety percent of all technical publications on the gator come from work in Louisiana." I had been given a stack of such publications, several by him. His studies and others like it were part of a program to bring back the alligator as a "renewable resource." By 1972 state biologists' censuses put the number of alligators in Louisiana at 250,000, and the decision was made to allow "an experimental harvest." Many conservationists, led by the National Audubon Society, protested.

Behind the scenes, members of the Louisiana Wild Life and Fisheries Commission met with officials of the Federal Government's Office of Endangered Species and International Activities, which administered the Endangered Species Conservation Act of 1969 and maintained the federal list of endangered species. Federal officials refused to take the gator off the list. Asked if they could prevent the harvest, they had to say that under existing law they could not.

"You've got to understand," an official told me privately, and bitterly. "We deal with 49 states—and Louisiana."

If Ted Joanen was bitter, he didn't show it. But in his briefing he said, "We're waist-deep in gators, as they say around here. There are so many of them they could inflict harm on other animals. Putting the alligator on the endangered species list was an overreaction."

At dawn, participants in the 1972 hunt set out across Lake Misere to check their meat-baited lines for gators. The alligator's broad snout and tooth pattern distinguish it from the American crocodile—an even rarer species—with snaggly exposed teeth and pointed snout.

American alligator

DAVID R. BRIDGE, NATIONAL GEOGRAPHIC STAFF (ABOVE AND TOP)

American crocodile

MARTY STOUFFER, NATIONAL AUDUBON SOCIETY

The harvest area excluded the refuge and covered about 278,000 acres of adjacent private land, where the alligator population was estimated at 50,000. Each hunter needed a special license and permission from a landowner to enter specific hunt areas. Each was issued metal tags, the quantity coinciding with the number of gators allotted to his area. The total quota was 4,000. "It's built into a trapper's way of life not to cross property lines," Ted told me. "So we know that the division of the marshland into areas will be honored. And there's rigid law enforcement, with good court backing. There's a mandatory jail sentence for poaching here in Cameron Parish — six months to a year, depending on the judge." The most recent poacher arrested was a repeater. When the hunt began he was serving five years, 165 days.

Ted Joanen's estimate — *50,000 alligators* — was the first thought I had early the next morning when I found myself sinking in the black water of Lake Misere marsh: opaque water, 50,000 pairs of jaws, ooze yielding to my wildly pedaling feet. Four of us — National Geographic colleague Dave Bridge, state enforcement agent James Collins, Frank Davis of the wildlife commission, and I — had been in a bateau being towed, at water-ski speed, by a vessel known as a "mud boat." The flat-bottom, 18-foot mud boat consisted mainly of an automobile engine and a huge auger seemingly big enough to drill for oil. What it drilled was the goo of the marsh — "three feet of mud and two feet of water," James said with the pride of a native as I slogged avidly toward the foundered bateau.

"Never had any trouble towing a boat with alligators in it," the grinning skipper of the mud boat told me. He had veered down a slough with a hard-to-port, and our light bateau had kept going straight, tipped me out as it skimmed up a bank, spilled my companions, and nosedived back into the water. We bailed our boat, shortened the towline, climbed in, and the mud boat roared off as fast as before, its wake as thick as chocolate-cake batter. The thin batter soaking my clothes baked in the hot sun.

The three hunters in the mud boat took an uncharted course through the marsh. At intervals the boat would nudge up to a bank where a nylon line hung from a branch or a bamboo pole. At one end of the line, a few inches above water, was a hook; the other end was securely tied on shore. Some hooks were bare, stripped of their hog- or cow-liver bait.

The first alligator they found was drowned, wrapped around the rotting pilings of one of the many abandoned oil-well heads — and working oil and gas wells — that dot the marshlands.

They got their second and last alligator of the day nearly two hours later. It thrashed in the water even after a charge of No. 1 buckshot blew a quarter-inch hole in its head. They hauled it aboard, severed its spinal cord, cut a slit in its tail,

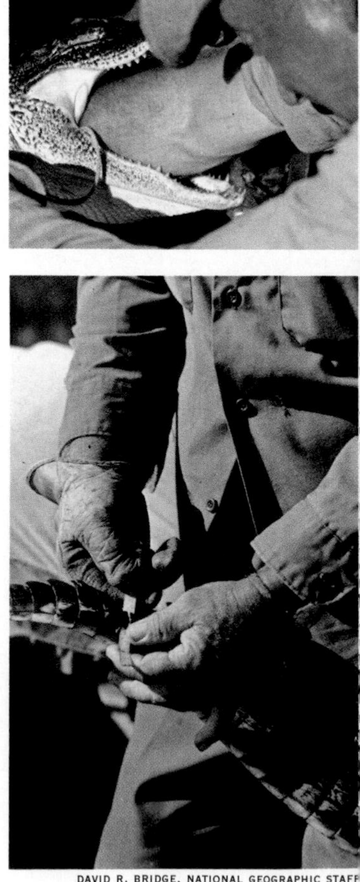

Thrashing struggles of a hooked gator will end with a single shot to the brain. After retrieving the swallowed lure from the stomach, the hunter marks his catch with a numbered brass tag. These labels and other measures enabled authorities to control the "harvesting."

DAVID R. BRIDGE, NATIONAL GEOGRAPHIC STAFF

DAVID R. BRIDGE, NATIONAL GEOGRAPHIC STAFF

Roaring along the bayous near Lake Misere, hunters return home with their catch lashed across the bow of a "mud boat." In the side yard at Bernell Koch's house, men strip the salable underbody and the legs of a gator, carefully avoiding cuts that lower the value of the hide. In 1972 much of the meat went to waste, although it might find use in stock feed—or a place on a family dinner table. "Gator meat's real tasty," volunteered one state wildlife official.

and inserted a numbered brass tag stamped LA. WILDLIFE & FISH COMM. LA. 72.

Now, in the shade of a chinaberry tree in Bernell Koch's side yard a little barefoot girl is standing on one of the hides lying around. She says she is measuring herself for alligator shoes. Behind her, ignored even by the dogs, sheer waste, is a pile of what the hides had been peeled from: pale, glistening chunks of meat that look like apparitions of alligators. As a rule only the underside is marketable, so each carcass still bears the emblems of the species—the dragon-like battlement of scutes, the blunt snout, the cowled eyes.

Ted Joanen, standing with me in Bernell's yard, pointed to a hide and ticked off the hands that it would pass through: buyer, dealer, tanner, cutter, manufacturer, wholesaler, retailer, owner of a pair of shoes or a handbag. "That hide," he said, "will feed a lot of families in a lot of places."

One of those families will certainly be Bernell's. I saw him run his 125 lines one day, with his 12-year-old son as an eager helper. He hauled only four gators into his mud boat that day. I hadn't been along the day before, when he caught 31.

They were calling him the champion when he brought his hides into the long metal shed at the refuge. The hunt was over now, but it would be a while before Bernell and the other hunters learned how much their harvest was worth. Ted Joanen, other biologists, the refuge staff, and enforcement agents were checking the hides for the brass tags—and for front feet. A surprise skinning technique had been ordered. Hides traditionally do not bear feet; any poacher who had slipped a pre-hunt "freezer hide" among the legal ones would have been caught. One hide, mistakenly cut by the old technique, was rejected. About 80 percent of the gators taken were males, confirming studies that showed the females would be nesting. The hunt, by Ted Joanen's biological standards, was a success.

Next day came the dealers who would decide whether the hunt had been commercially successful. Each presided over a long table where, one by one, the hides were unrolled, measured, graded, and then re-rolled. They would be sold eventually at auction; prices would average $55.93 per hide (as against approximately $20 in 1963). Somewhere in the rolls of hides was Old Bread's. He was 11 feet 7 inches long, and if he sold for the highest price, he brought his killer at least $120.

While I watched the sweating, unsmiling dealers at their work, an enforcement agent came up to me and nodded toward the busiest table, overseen by a young man wearing a baseball cap. "That's Chris Plott," he said. He was not the only state agent staring at Chris Plott, subject of a federal case.

Even some Louisiana officials had been surprised to learn—as was I—that federal law did not forbid the killing of animals on the endangered species list. "Right now," one biologist told me in 1972, "you can go to Texas and shoot every red wolf you see and not violate U. S. law. You could kill any endangered species—except those protected by other statutes." These give special protection to migratory birds, to eagles, to marine

DAVID R. BRIDGE, NATIONAL GEOGRAPHIC STAFF

In a shed on Louisiana's Rockefeller Wildlife Refuge, tubs of salt-preserved hides await the auctioneer's gavel: 1,337 legally taken skins will change hands. Representing his family's fur-and-skins export business, buyer Christopher Plott checks the wares before bidding starts. He attended the sale before serving a 60-day prison sentence for illegally shipping alligator hides in foreign commerce—to Yokohama, Japan. Although laws prohibit the killing of gators, lucrative markets abroad have provided a constant incentive to poachers. One underground network, broken up in 1971, had marketed 127,000 alligators valued at more than $4,100,000.

mammals, and to any animal taken in violation of a state law and then moved across state or international borders.

Before December 1973, a state had sovereign power over its wild animal residents. Since state laws protected the alligator throughout its range, the Federal Government had rarely invoked its interstate-traffic laws against poachers. One of the first such actions was brought six months before the Louisiana hunt: Quince C. Plott was charged in New York with shipping 4,788 poached alligator hides from Georgia. He and his son, Christopher, were also indicted in Georgia. The father pleaded guilty to 67 counts, the son to 66.

The elder Plott was sentenced to six months in prison. Christopher had not yet begun his 60-day sentence when he was pointed out to me in the shed. His father's term had started a few weeks earlier. The sentences had been arranged so that the family export business could go on without interruption.

Evidence gathered in this case vividly revealed the potential fate of a "protected"—but economically valuable—animal. Federal investigators found records indicating the marketing of about 127,000 hides between 1968 and 1971; estimates of the total killed in that period run as high as 500,000. Since 1970 a New York state law, the Mason Act, has prohibited the sale of any product made not only from alligator but also from other endangered species. California, Connecticut, Massachusetts, Illinois, and Pennsylvania have similar laws. But closing the lucrative American markets did not stop the trade. Legal hides of the Louisiana harvest would follow the overseas route of poachers' hides: to France to be made into purses and luggage, to Japan to be made into belts and wallets.

If the alligator can survive despite such slaughter, and if its estimated million individuals actually exist, why is it on the endangered species list? Because large numbers do not necessarily guarantee survival. There were so many passenger pigeons in 1869 that 11,880,000 were taken for the squab market from one Michigan nesting area in one 40-day period. By 1900 the species was extinct in the wild.

Listing does not necessarily mean that numbers are down to a critical level—quite possibly no one knows what that level is. Nor is the list more than a starting point for finding out how many North American animals are in trouble, officially or otherwise. The official list appears in the *Federal Register*, the daily publication that gives executive orders and regulations the force of law. Under the Endangered Species Act of 1973, the Secretary of the Interior must publish on appropriate occasions the names of native species that he decides are "in danger of extinction"—or, a new category, "threatened," likely to become endangered within "the foreseeable future" in "all or a significant portion of its range." He makes his decisions "after consultation, as appropriate, with the affected States" and interested parties—and must publish his reasons.

The law now gives him power to protect these species. Like its predecessors, it deliberately leaves hazy the requirements for listing—which is one reason the list has fluctuated.

FREDERICK KENT TRUSLOW

In Everglades National Park, a 12-foot bull gator lumbers into the water with scarcely a ripple; few of the species grow so large today. Florida law forbids the killing of the reptile. Its future in this state hinges on that of its environment, currently at risk from the reclamation of marshes that drain into the Everglades.

As of 1973, there were 109 species—or subspecies—on the list. Of these, 29 are birds of Hawaii or Puerto Rico, one is a Hawaiian bat, one is a Puerto Rican snake. Theoretically, then, only 78 North American species or subspecies are endangered. But scores of others are considered equally endangered by experts who have not managed to get "their" species listed. Canada and Mexico also have imperiled species—but neither has an official list. Canadian authorities I talked with count 66 endangered animals, some of which have ranges that overlap the border and some of which are *not* on the U. S. list. At least 10 animals on the U. S. list also range into Mexico.

Adding to the confusion is the 289-page *Threatened Wildlife of the United States,* the so-called "red book" compiled by the Office of Endangered Species. Between red-for-danger covers are many species that, at first glance, would appear to be officially endangered. But after the 1973 edition was published an "errata" sheet was distributed. This made inoperative all remarks about species except those in the *Federal Register,* and thus it left 69 animals in a kind of federal limbo.

Some states have lists that claim endangered status for animals not federally recognized. Even the Federal Government has conflicting lists. The Forest Service, an agency of the Department of Agriculture, describes as "unique," "rare," or "endangered" several species not recognized as threatened by the Department of the Interior's Office of Endangered Species.

Tracking wildlife through a governmental maze, I learned that no fewer than 550 federal programs affected animals ranging from jellyfish (Bureau of Commercial Fisheries) and mosquitoes (Public Health Service) to whales (National Oceanic and Atmospheric Administration) and bears (National Park Service). The Department of Defense in 1972 ordered 277,502 parka hoods trimmed with wolf fur, though the red wolf and eastern timber wolf were on the O.E.S. list. It canceled the order after conservationists and Congressman G. William Whitehurst of Virginia protested; Interior pointed out that the amount of fur involved might affect all the gray wolves of the continent.

Down the corridors of the Department of the Interior one trail may lead to the Office of Endangered Species but another may end at the Division of Wildlife Services. Its primary service is killing animals labeled as pests, including wolves and mountain lions, which many private animal-protection organizations consider among our most imperiled species.

I searched in vain for an ark. There is no ark this time. Saving our wildlife will take more than loading them, animal by animal, aboard a list or a program. "Most of the animals on our list," says Gene Ruhr of the Office of Endangered Species, "are challenged by environmental threats. Our ability to prevent the loss of nature's animals is an indication of how we are managing our own environment. If we could recognize what is best for man we probably would not need an endangered species list."

We long have portrayed ourselves as members of the supreme species that reigns from the summit of the mountain of life. Around us, but slightly lower, *(Continued on page 56)*

Beyond recall? Beyond the law? A portfolio of endangered species

For one, any measure comes too late; for others, survival may turn on international agreement to control a luxury trade or establish a law of the sea.

Passenger pigeons alight at the edge of a forest in search of pinecone seeds. In the 1800's hunters shot these birds in millions, unaware that the highly social pigeon required the stimulus of large flocks to breed successfully. The species had become extinct in the wild by 1900.

Following the Gulf Stream, an Atlantic loggerhead journeys to nesting grounds on North American beaches between the Florida Keys and North Carolina. Now proposed for the federal list of endangered species, it faces a tenuous future. Predation, especially by raccoons, destroys thousands of eggs annually. Of turtles hatched, few survive the trek seaward. Lights from beachside developments draw them inland—to end up smashed on highways or baked to death in the sun.

Overleaf: With fluttering wings, an Eskimo curlew courts his mate on the Canadian tundra—a pose the artist took from a related species of known habits. Traveling more than 8,000 miles from its breeding grounds to its winter home in Argentina, this curlew fell victim to guns that lined the flyways during the 19th century. Listed as endangered, it has all but vanished. Only eight recorded sightings since 1950 witness its lingering existence.

Claws raised, male American lobsters engage in a territorial display. Overharvesting and pollution have depleted much of the inshore supply along the Canadian-New England coast. Although the federal law provides for listing of crustaceans, the government does not consider the species threatened.

After a vigorous tail-beating motion, a paiute cutthroat trout deposits her eggs; attending males will fertilize them. Found in two California counties, this race faces extinction by overfishing and hybridization. Hoping to save the trout, an angler's favorite, the state has begun recovery programs; the Federal Government lists the subspecies as endangered.

Rearing in the grass, a San Joaquin kit fox pounces on a kangaroo rat. Dwindling habitat, poisoning from rodenticides, and illegal hunting have brought the fox's numbers below 3,000. Subject of state wildlife studies, it figures on the federal list.

Probing the mud with its barbels, a 50-pound lake sturgeon forages for mollusks. Once abundant in the Great Lakes, in 1885 it yielded a catch of 8,500,000 pounds—as against 1,000 to 3,000 pounds in recent years. It has no listing.

Inch by inch, a jaguar stalks its prey in the Yuma Desert. This large carnivore ranges from Argentina to Mexico, with some stragglers reaching Texas and Arizona. Prized by furriers abroad, the jaguar arouses international concern due to its fast-decreasing numbers. Although Mexican law limits hunting and bans all export of skins, shooting (for sport, protection of livestock, or smuggling) and loss of habitat have reduced the jaguars within its borders to about a thousand.

passenger pigeon

PAINTINGS BY JAY H. MATTERNES

Atlantic loggerhead turtle

Eskimo curlew
American lobster
paiute cutthroat trout

San Joaquin kit fox
lake sturgeon
jaguar

Bill gaping wide with the lower mandible expanded, a brown pelican inverts its pouch and exposes the structures of the throat. The activity—technically known as a "comfort movement"—lasts from two to four seconds and usually occurs during preening.

JESSIE O'CONNELL GIBBS

JEN AND DES BARTLETT (ABOVE, UPPER LEFT, AND BELOW)

On a 1970 trip to Isla las Ánimas in the Gulf of California, federal biologists from Denver spray dye on a brown pelican to trace its travels for a long-range research project, still under way. They want to determine if birds that migrate from five Mexican breeding colonies into California accumulate DDT and its derivative DDE by eating contaminated fish off the West Coast. Scientists have linked the pesticides to a reproductive failure in this officially endangered species: the laying of thin-shelled eggs that break or never hatch.

At left, a chick about six days old huddles by a menhaden; a parent will partially digest such fish, then regurgitate it onto the floor of the nest for the chick. Assuming a more active role, a five-week-old nestling plunges its own beak into its parent's gullet, forcing her to disgorge food.

are our fellow mammals, great and small. The birds soar and flutter for our delight. Fishes, especially those that rise to our hooks, are considered acceptable. Snakes, frogs, alligators? Yes, reptiles and amphibians are there. But most of them figure as villains in our myths. At this point the vertebrate zone of Life Mountain ends. Far below, disappearing from sight, are the invertebrates: insects and worms, sponges and jellyfish, snails and squids. We may enjoy eating a lobster or seeing a butterfly, but from our lofty view we give little heed to the creatures we call the lower life forms.

The endangered animals usually given official status reflect this view from the mountain top. The federal list of October 1973 names 17 species or subspecies of mammals, 53 birds, 31 fishes, 8 reptiles and amphibians. It stops there. The freshwater mollusks, a great treasure, are left out. Specialists consider 325 species of them endangered—and ignored. The patriotically named *Homarus americanus,* the lobster disappearing from United States and Canadian waters, is not on the list.

Robert Rush Miller of the University of Michigan, as chairman of the endangered species committee of the American Fisheries Society, reported in 1972 that 305 kinds of freshwater fishes in the United States were threatened. I thought of Stephen Vincent Benét's lines—*"I have fallen in love with American names,/The sharp names that never get fat"*—as I scanned Dr. Miller's list. Something more than a fish will be gone if we lose the mooneye, the popeye shiner, the bigeye jumprock, the western tonguetied minnow, the toothless blindcat, the pumpkinseed, or the Tippecanoe darter.

I was discussing the problem of formal lists one day with several curators of Canada's National Museum of Natural Sciences. Dr. E. L. Bousfield, the museum's chief zoologist, happened by and joined in. "Probably many hundreds of species unknown to science are actually endangered or extinct," he said. "It's sobering to consider what may be going on that we'll never know about. For example, the freshwater jellyfish may form a 'bloom' in a lake some year and then not be seen again in that lake for 25 years. What causes that kind of cycle?"

He sighed. "But society at large probably doesn't care because of its concern with the more conspicuous and economically important animals." He brought his hands together, touching fingertips. Then he dropped his hands to his sides and the image of a mountain vanished. "Someday, we may find that the food pyramid has collapsed—that's when it will hurt us."

We spend most of our wildlife resources—money, personnel, land, research—managing a certain group of those important animals high on the mountain and the food pyramid—the game species. We can keep them off the endangered list.

The game species seem to be flourishing. The birds among them are counted in the millions and tens of millions. But the remark of a longtime hunter in Louisiana aroused my curiosity about one particular group—the waterfowl of North America. "There aren't as many geese around as there used to be," he said. Where, he wondered, had all the geese gone?

GRANT HEILMAN (BELOW); JEN AND DES BARTLETT

Beside a shrimp boat off Mexico's Isla Pelícano, namesake birds vie for scraps while hovering gulls wait to filch a morsel. Here, in the south, brown pelicans still do fairly well. But at California's sole breeding colony, on Anacapa Island, nesting attempts have dropped from 1,272 in 1969 to 247 in 1973. Conditions vary for the eastern brown pelican; only remnant groups inhabit the Gulf Coast. Along the Atlantic, trouble seems worst in the north, decreasing southward. At left, a bird in its winter plumage perches on a pier in St. Petersburg, Florida.

3. The Gunners' Targets: Game Species and Game Harvest

Protected under law, imperiled nevertheless, the hunted face new risks in a bio-political world

The eight of us—six hunters, a Chesapeake Bay retriever named Josie, and I—stood by the blind that hunched on the shore of Southeast Creek. We were awaiting the sunrise and my host Mervin Cohey was anticipating "a good bad day," the kind of raw, blustering January day so miserably admired by the duck hunters of Maryland's Eastern Shore. I am not a duck hunter. I was contemplating my feet, my freezing feet, and the darkness, the freezing darkness.

Then dawn drifted in on the chill wind, and I could see black dots strung down the middle of the creek: Canada geese, hundreds of them, honking irregularly, drifting almost imperceptibly in long, rippling rafts. Amidst the honking I could hear an occasional quack of an unseen duck, the hooting of invisible swans. I saw the swans only for a long moment as they pulled themselves into the air and vanished in the fog.

In twos, and threes, and fours, the geese began to fly up and vanish, too. "The honking will stop, all of a sudden," Mervin whispered. "There won't be a single leftover honk."

In a moment there was not a sound. In another moment from the farther shore I heard the muffled *pop pop pop* of hunters'

Safe for the moment, a giant Canada goose in Sand Lake National Wildlife Refuge, South Dakota, guards her five fuzzy goslings. Hunting and habitat loss brought this subspecies near extinction by the 1920's.

JEN AND DES BARTLETT

Southbound birds signal the arrival of fall—and an open season on many waterfowl. Following age-old flight patterns to warmer wintering grounds, migrants from the north funnel into four main corridor zones: the Atlantic, Mississippi, Central, and Pacific flyways. Treaties with Mexico and Canada ensure their passage. The United States has an administrative region for each flyway, with a council of state experts, and a national council to help regulate yearly waterfowl supply.

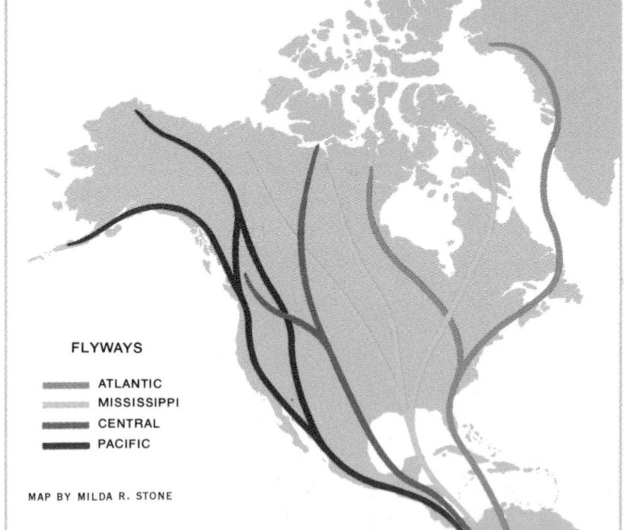

NATIONAL GEOGRAPHIC PHOTOGRAPHER JAMES L. STANFIELD (ABOVE AND LOWER LEFT)

A flock of Canada geese clatters from corn planted in Horseshoe Lake Wildlife Refuge, Illinois, a "lure crop" to draw them away from farmers' fields. At left, a conservation officer walks with hunters after checking geese they bagged on state-owned land adjoining the refuge. Grain crops in northern states may interrupt or halt the southward course of migrating waterfowl. Game managers call this "short-stopping"; it provokes sharp controversy, especially in southern states.

guns. Then utter silence, instinctive self-protection of prey.

Bobbing in front of the blind, arrayed separately, were sets of goose and mallard decoys, each a work of art. Mervin hoped that the wind and cold of his good bad day would drive birds out of Chesapeake Bay, into sheltering creeks and marshy sloughs—and then, if he had luck, to his blind.

A string of black ducks appeared, and one peeled off. Guns went up. But no one fired. It was not what Mervin calls a good shot. He will not shoot at a bird beyond 35 yards, and the bird must be "tolling," a hunter's word for that marvelous gliding flight when a bird locks its wings and sweeps down from the sky.

The creek was a gray slate now, wiped clean of geese. They were aloft in skeins that frayed when a dozen or so birds broke off, formed into ragged V's, and, faintly honking, soared into the fog. As they rose, they became gray ghosts, seemingly transparent. I was still cold but I did not want more light. I wanted more moments like these, when the geese and the sky were one.

Only when a goose wheeled down and tolled past the blind did I remember why we were here. Mervin raised his gun. Too late. The goose sailed by. "The boys on the point should have got him," he said. "It was a good shot."

A solitary black duck flashed by. Six guns rose, swung, and popped. The duck wavered, its wings fluttering. Then, like a black leaf falling on a still day, it spiraled slowly down. Mervin's nephew and a friend quickly put out in a small boat and killed the duck in the water.

The next bird, second and last of a 5½-hour day, was a mallard. It also fell on the far side of the creek. This time when the boatmen headed back they dropped the duck about 40 feet from shore. Mervin sent Josie to fetch it. If a Chesapeake Bay retriever can smile, Josie did. She slipped into the water, clutched the bird in her mouth on the first try, paddled back, and deposited the prize at her master's feet. Mervin looked at his friend George Founds and nodded. "Good dog." George had acquired Josie in a trade of Canada geese and then had given her to Mervin, from whose blind George had shot the geese.

Once such neighborliness was as much a part of the hunting tradition of the Eastern Shore as weather-wise hunters, shooting etiquette, and hand-carved decoys. Today a machine can produce a dozen decoys at a time, farmers like Mervin complain about trespassing strangers, and Maryland's fraternity of Chesapeake hunters has become an army of 40,000.

I had witnessed on Southeast Creek something fast disappearing: "quality hunting," waterfowl managers call it.

Federal regulation of migratory bird hunting really dates from 1918, two years after the signing of the treaty to protect the Canada-United States migrants. Until 1929, seasons ran for about 100 days and the daily bag limit for each of the 500,000 hunters was 25 ducks, 8 geese, 8 brant, and 25 coots. By the 1960's the seasons had ebbed; the bag limit was down to 2 to 6 ducks and geese; the number of hunters had passed 2,000,000. By 1971 some of the finest game birds of the past were in trouble.

Today the bag limit hovers at around 5 ducks, 5 geese, 15

NATIONAL GEOGRAPHIC PHOTOGRAPHER OTIS IMBODEN

In modified turkey crates, Aleutian Canada geese raised at the Patuxent Wildlife Research Center in Maryland arrive at Amchitka Island. Released, the birds tentatively inspect a frozen pond on their new home. Arctic foxes introduced for their fur had reduced the race to a small colony on remote Buldir Island; scientists cleared Amchitka of foxes before trying to restore geese to their former breeding grounds. Despite all precautions, the ex-captives disappeared six weeks later; their fate remains unknown.

coots. Hardly a state has a simple, date-to-date season. Federal regulations are studded with footnotes of exceptions. This new, bewildering complexity reflects the arcane art of waterfowl managers who must raise their bird crops on steadily shrinking lands, supervise steadily increasing harvests—and guard against the ultimate over-harvest: extinction.

"If every licensed waterfowl hunter got his bag limit on just five days of the season," says Robert Arbib, editor of *American Birds*, "the entire waterfowl population would be wiped out." Hunters don't shoot that well and the regulations allow for the fact, but in a recent year 2,398,350 of them did bag 17,108,700 ducks and 1,752,400 geese. An estimated 50,000,000 waterfowl survived the harvest.

Some biologists worry about such kills. Waterfowl managers, however, proclaim as an article of faith that legal hunting is not a threat to any species in this country.

But these same managers readily admit that waterfowl encounter many other perils. Take that Canada goose which swooped safely past Mervin Cohey's gun. It may evade every hunter and yet succumb to their shooting.

I went to Centreville, Maryland, not far from Mervin's blind, to learn how hunters can miss a bird and kill it. V's of geese were scissoring the sky that December day, and the high school auditorium in Centreville was buzzing with talk of ducks and geese. The meeting, co-sponsored by the National Rifle Association and Maryland Ducks Unlimited, had been called by the state's Department of Natural Resources to inform the public about a silent killer, lead poisoning.

Vernon Stotts, waterfowl specialist of the Maryland Wildlife Administration, stood on the stage and gave the past season's toll of dead geese found on the Delmarva Peninsula. In one area, 513 . . . 696 here . . . 650 there. And some that were dying: "There was a definite vile odor. . . . Often the eyelids were swollen and encrusted. . . . They were easily captured. . . ."

Then came Dr. Louis N. Locke of the U. S. Fish and Wildlife Service to tell how and why they died: "There is a muscular weakness. Next the bird loses the ability to fly. Frequently the esophagus is impacted—stuffed with sand, corn, and mud. The pectoral muscles waste away. Anemia sets in. Sometimes when you pick up one of these emaciated birds, it struggles, suffers an acute cardiac attack, and it dies." He told what happened inside the bird: "muscles, tissues, nerves are not getting their normal nourishment of blood, and they die. . . . Lead injures the kidney. . . . Lead kills the cells of the heart. . . ."

About 5,000 geese died that way in Maryland, some of the estimated 2,000,000 geese, ducks, and swans that die each year of lead poisoning by ingesting shotgun pellets. Hunters fire about 4,700 *tons* of pellets each year. The pellets must feel like seeds and gravel to a bird as it probes lake and marsh bottoms for food and grit. If enough poisoned birds die at one time, the bodies are found. No records tell how many die unseen or waste away to become easy targets.

Lead poisoning in waterfowl has been recognized since 1842,

and Frank C. Bellrose of the Illinois Natural History Survey has been writing scientific reports on it since 1947. His work left no doubt that spent lead pellets were poisoning waterfowl. Biologically, there are ways to stop this; one is to substitute iron shot for lead. Iron can be harmlessly absorbed, as demonstrated by feeding iron pellets to penned wild mallards.

But the decision to use iron shot would not be made by biologists. It would involve economics and politics. A changeover would be expensive, said manufacturers of ammunition. And hunters wondered whether iron shot would harm their guns or impair their shooting. State and federal officials had to weigh the poisoning against these considerations.

In 1973 Maryland became the first state to ban the use of lead shot on waterfowl. A federal official announced that the Fish and Wildlife Service intended to move as rapidly as possible toward a nationwide ban on the use of lead shot.

The lead-poisoning dilemma illustrates how biology and politics can converge on a wildlife problem. The phenomenon has added a word to the language: bio-politics.

Our Canada goose, which flew past Mervin's gun and perhaps escaped lead poisoning, exists in a bio-political world with hazards and sustenance as real as those of its natural world. Hatched on public land in northeastern Canada, welcomed to state and federal refuges on its wintering grounds, it gets many a meal from men—some of them patrons on government payrolls, some of them irate farmers. Banded, counted, scanned by radar during migration, it lives under relentless human surveillance and is managed from the egg to the bag.

Though targets of hunters, Canada geese and the other waterfowl of North America are beneficiaries of protective laws and land-acquisition programs, both private and public, which help to preserve imperiled habitats. Under long-honored treaties, migratory birds are the wards of the United States, Canada, and Mexico—and as of 1973, Japan. Waterfowl share this status with less spectacular migrants, including sparrows, and with warblers and vireos, hummingbirds and tanagers. The United States treaties, dating from 1916 for Canada, from 1936 for Mexico, cover hundreds of non-game species.

Treaties and laws staked out the waterfowl's bio-political habitat. State and federal game management works to sustain the habitat, producing the annual crops of ducks and geese for the reapers of fall. The fate of North America's 48 waterfowl species turns on how that habitat is farmed, how those crops are reaped. I asked a biologist what percentage of waterfowl live in the bio-political world of game management. "I'd guess about 95 percent," he said. And when I asked what would happen if management suddenly stopped, the biologist shook his head and said, "I guess that would be the end."

Only two of our waterfowl species do not have extensive migrations: the mottled duck of the Gulf Coast and the Mexican duck, found from northern New Mexico to central Mexico—

and on the endangered species list. Many other waterfowl migrants might be on that list today—or perhaps extinct—if in the 1930's they had been left to their collapsing natural world.

The plowed-up prairies were shriveling under a drought so terrible that it created the huge, devastated area called the Dust Bowl. On the prairies of south-central Canada and the northern tier of the central United States, most of the continent's ducks were breeding. Pairs of ducks sought out nesting sites near glacier-made ponds or basins called potholes. Now the drought —and the draining of pothole lands for agriculture—was threatening the heartland of the breeding grounds. If the birds were to be saved, the land had to be saved. But how? The United States Government, in the midst of the Depression, was not buying much land. The Canadian Government had no thought of purchasing waterfowl habitat.

Rescue measures began with the passage of the Migratory Bird Hunting Stamp Act in 1934 and by the founding of Ducks Unlimited in 1937 (in Canada a year later). The private, non-profit organization began raising funds, and in 1938 was supervising an area in Manitoba's Big Grass Marsh, within a region called "the duck factory." Ducks Unlimited today operates more than 1,000 Canadian water-management projects that provide habitats for a variety of creatures.

In the United States, the act that became known as the Duck Stamp Law launched a land boom for waterfowl. To qualify as "a legal wildfowler," every hunter 16 or older had to have not only a state license but also a $1 federal duck stamp. Each dollar went into a special fund, 90 percent of which the Fish and Wildlife Service had to spend on refuges for waterfowl. The other 10 percent could be spent on law enforcement and on the stamps. Designed by unpaid artists in annual competitions, the stamps unexpectedly attracted philatelists. Hunters accepted them just as avidly. The Duck Stamp Law was followed by others that slowed down the conversion of wetlands into croplands, aided states in the acquisition of habitat lands, and opened national wildlife refuges to hunting.

Managers were growing wiser to the ways of waterfowl. Studies, spurred by the breeding-ground crisis, showed that the birds' traditional nesting demands were difficult to change. As they migrated, however, they seemed to lose their conservative habits and could be induced to try new wintering places.

The time-honored view of the waterfowl flyways has changed. Now managers perceive each flyway as a complex of flight paths and corridors, along which migrations can be predicted—and perhaps altered. Each flyway—Atlantic, Mississippi, Central, and Pacific—has become an administrative region monitored by a council of state experts and advised by a U. S. Fish and Wildlife Service coordinator. Representatives from the flyways form the National Waterfowl Council, whose control over waterfowl is roughly comparable to the Federal Reserve Board's control of the nation's money supply.

I met one of the continent's authorities on migration and waterfowl management, Frank C. Bellrose, at the meeting on

MICHAEL S. SAMPLE

Necks thrust forward, trumpeter swans fly low in Yellowstone. Breeding colonies—established in the national wildlife refuges of Malheur, Oregon; Ruby Lake, Nevada; Lacreek, South Dakota—have begun to re-extend the trumpeter's badly reduced range.

Below, rusty-headed whistling swans tend their young cygnet; one preens its snowy feathers. Unlike trumpeters, the whistlers no longer enjoy complete protection; with their numbers increasing, Utah, Montana, and Nevada now allow short hunting seasons.

JEN AND DES BARTLETT (ABOVE AND LEFT)

lead-poisoning. "More than any other species of waterfowl," he had written in 1968, "Canada geese have radically altered their migration routes. . . ." This he explained by their "rapid adaptation to newly created waterfowl refuges and feeding grounds." I was curious about the way the geese were being changed. But when I talked to Mr. Bellrose, I found he was more concerned about great changes being made on the face of the land. He suggested I go to the Mississippi Flyway and see for myself.

We skimmed about 200 feet above the Illinois River. Then pilot Bill Houlihan rolled the light plane and I was once again looking straight down on a black hole in the ice surrounded by a rippling mass of birds. How could there be so many? Flock after flock, mile after mile. The temperature had been 8 below when we took off from the snow-laced strip near Manito, Illinois. And I had wondered if we would see any birds on this December day. Now "Tud" Crompton of the Illinois Natural History Survey was counting again, and Gary Senn, a staff biologist of the state's Department of Conservation, was jotting the incredible numbers on his census sheet.

"Mallards . . . 49,000." Another few miles, another black hole in the ice: "Mallards . . . 67,000 . . . Canada geese . . . 450 . . . lesser scaups . . . 250 . . . There's a bald eagle, a juvenile. . . ."

Between counts, his eyes on the river unrolling below, Tud explained how they did it. "We've taken photographs and counted the birds in a given area. We figure out the dimensions of the area, and from that we can figure the number of birds that will fill it. The average situation is about one square yard per duck. We're about 90 percent accurate." And the species identification? "It's easy," Tud said. "You just get to know, by long association. How they fly, how they act. That eagle, he looked like a juvenile. And anyway, if he was an adult, he'd have those mallards a lot more worried."

The Mississippi, a mosaic of ice blocks and dark water, sustained dozens of aggregations of swirling ducks. Each flock wheeled around a hole in the ice, kept open by hardy swimmers; the bigger the hole, the bigger the flock. The mallards, tardy migrants to southern warmth, will tarry as long as they can find open water and food—remnants of crops on nearby farmlands, or standing grain on nearby refuges.

Lack of food, not the cold, drives the birds south, Mr. Bellrose later told me. Apparently the mallards found the food plentiful. On a freezing mid-December day, the mallard count along 100 miles of the Illinois River was 256,990, and along 150 miles of the Mississippi it was 104,700. With numbers like those, how could the mallards be in trouble? "Their main wintering ground is in Arkansas," Mr. Bellrose said. "That's where the trouble is, along the Cache River." I would migrate farther south myself, for a bird's eye view of the Cache.

Now, though, I was grounded near the river town of Havana, Illinois, learning about missing clams and snails. In the Natural History Survey laboratory, Dr. Richard Sparks was building a small tank in which he would attempt to re-create the riverbed realm of fingernail clams.

Here in Frank Bellrose's headquarters I could see why his beloved Illinois River has been called the most studied river in the world. So complete are the biological records going back to 1903 that Survey scientists can plot, mile by mile, the relentless advance of pollution from Chicago to Havana, 207 miles downstream. And Frank can point to a graph that shows how, for want of a clam, a duck was lost.

Yearly from 1946 until 1954 the river attracted an average of 1,000,000 lesser scaups, the little, blue-billed ducks that dive to feed on clams and snails. Since 1955 the lesser scaup has virtually vanished from the river—as have the fingernail clams and the little snail impressively named *Cincinnatia emarginata*. The number of ring-necked ducks also dropped, but not as drastically, probably because they were able to supplement their diet with aquatic vegetation.

The tank being built in the laboratory seemed pitifully small for such a vast restoration task. But at least it was a start toward trying to learn why the clams and snails had vanished from what Thomas Jefferson called "a fine river, clear, gentle," and a visitor in 1838 described as "infested with wild beasts."

Human intervention in a bird's way of life is not usually so sophisticated as this research. Nor is management restricted to refuges. Farms are favorite lunch stops on migration routes. Frank once saw mallards sweeping into a 20-acre cornfield at the rate of 1,000 a minute. Canada geese can eat through a field by such sleight of bill that at a distance a farmer would not know he had been burgled. The goose, which can stretch to get corn 3½ feet up a stalk, sticks its head into a husk and strips the ear, leaving the husk apparently intact.

In some provinces and states the damage costs are in the millions, and public funds are used to compensate the bird-blitzed farmers. A 1970 survey in 24 major corn-producing states showed that 6,367,201 bushels of corn were eaten, trampled, or contaminated by birds; they consumed about one-sixth of one percent of the entire United States corn crop.

Farming recently has also been hurting waterfowl where they live. Hundreds of thousands of acres of wetlands—prime wintering grounds—have been converted into fields for soybeans, the nation's new big-money crop. "Waterfowl's worst enemy," a Louisiana biologist told me, "is the soybean."

I saw miles of proposed new soybean fields when I once again looked down on a river—this one the Cache, which flows in great S-shaped curves from southern Missouri into eastern Arkansas. I was flying with a pilot-biologist, David M. Donaldson, chief waterfowl specialist of the Arkansas Game and Fish Commission.

As I saw the meandering Cache of today, he talked despairingly of the Cache of tomorrow. He was envisioning the results of a plan to channelize the Cache, a transformation of the winding river into straight ditch 140 miles long.

In my eyes, rafts of mallards on flooded forests, tipping for fallen acorns. In his, the wipe-out of yet one more of the

Under man's supervision, the waterfowl in the United States live and die. At right, mallards scramble for grain tossed into a pond for trumpeter swans in Malheur refuge. Below, at Bear River refuge in Utah, a wildlife aide collects ducks sick with botulism, hoping injections of serum will save them. A bacterial food poisoning, botulism produces paralysis and death. With habitat loss—and short-stopping—birds must concentrate in small areas where outbreaks of disease can take high tolls. In 1967, 37,000 ducks, mostly pintails, died in and around Bear River.

RON KLATASKE (BELOW) AND DAVID B. MARSHALL (OPPOSITE)

JEN AND DES BARTLETT (ABOVE AND LEFT)

At right, a cannon net fired by remote control traps ducks for banding. Yearly U. S. and Canadian agencies band some 300,000 waterfowl, study mortality rates with the aid of computers, and use the data to regulate their harvest. From a lakeside blind in Wyoming, a hunter takes aim; a Canada goose decoy sits in the short prairie grass behind him.

continent's last great mallard wintering areas, the gathering place for 250,000 to 300,000 mallards.

A broad swath of hardwood forest, interlaced with glints of water overflowing from the Cache.... The end of the classic Arkansas duck hunt, with decoys arrayed in the shallow water, a hunter in hip boots leaning against an oak.... The end of a cycling creation of habitat by short-term flooding, long-term forest growth....

Farm fields tapering off to timber stands that fill the S's of the Cache's graceful curves.... Thousands of acres of soybean fields marching to the straight-edged gutter that was a river, joining the 2,000,000 acres in Arkansas cleared of hardwood for farming in the past 20 years.

Just upriver from Clarendon, Arkansas, our views suddenly joined. A huge, orange earthmover stood on the stripped bank of the Cache. From a long black boom hung the basic tool of river-straightening, the dragline that scrapes the bank bare. It was meant to channelize 140 miles of the Cache, 77 miles of its main tributary, Bayou DeView, and 15 miles of small streams.

The Cache was still awaiting its fate when I flew over it. The project was only a few miles under way when it was stopped by court action. An environmentalist group had won an injunction. But the U. S. Army Corps of Engineers was continuing the legal battle. The Corps says that by taming the meandering Cache River-Bayou DeView system engineers will ease flooding on existing farmlands. Dr. Rex Hancock of Stuttgart, Arkansas, a dentist who is president of the citizens' group, says the project is "an insult to God's planning of the earth." The *Arkansas Gazette* said in an editorial: "In the long view, this state cannot live by soybeans alone." In a flood tide of controversy, the Cache winds on to its destiny.

NORMAN R. LIGHTFOOT, PHOTO RESEARCHERS, INC.

canvasback

JEN AND DES BARTLETT, BRUCE COLEMAN INC.

lesser scaup

Losing their age-old wetlands, waterfowl are also losing a more recent food source: farmers' leftovers. Waste grains once littered the fields after harvest. But improved mechanical harvesting leaves them less and less. Increasingly they depend on managers. On a flight over the Mississippi, I saw 100,000 Canada geese in the Union County refuge, operated by the State of Illinois. They were eating corn planted for them and, though tradition might have tugged them southward to warmth, full bellies kept them in frigid Illinois.

Waterfowl managers have a word for this: "short-stopping," the accidental or deliberate holding of migratory game birds in a state so that their southern migration is slowed down or stopped. As a result, hunters in short-stopping states have more geese to hunt than fellow sportsmen to the south.

That hunter in Louisiana who wondered where all the geese had gone apparently did not know about short-stopping. But waterfowl specialists certainly do know. And they were more than willing to talk about it, though mostly off-the-record.

One of them explained the waterfowl-hunting system to me, informally: "Every August the Federal Government sets up the regulations for hunting. The season framework is set from

RUSS KINNE, PHOTO RESEARCHERS, INC.

shoveler

KENNETH W. FINK, BRUCE COLEMAN INC.

redhead

N.G.S. PHOTOGRAPHER JAMES L. STANFIELD

wood duck

October 1st to January 20th. Then the Feds say to the states, 'You can take any 40 days, 70 days, and so forth, within that framework. You can have a point system or a bag system. You can split the season within the state.' Now state game commissions are highly political. Most game commissioners lack tenure, and many biologists do, too. If you are a biologist, your job may depend on how good you are at figuring the best way to hold the birds in your state during your hunting season."

Short-stopping is usually not discussed in public by state or federal officials. Much of what I learned came from concerned biologists who spoke in confidence and showed me unpublished reports that generally circulate only among members of the waterfowl managers' fraternity. But in recent years the subject has surfaced, particularly in the writings of state biologists in Louisiana. There in December 1970 fewer than 1,500 Canada geese were counted on wintering grounds that once attracted 100,000. In 1971, when biologists were convinced the same phenomenon was affecting blue and snow geese, *Louisiana Conservationist,* a magazine published by the Wild Life and Fisheries Commission, boldly stated the controversy:

"Many of the 900,000 blue and snow geese that winter along the Gulf Coast of Mexico, Texas, and Louisiana are being short-stopped on three small federal waterfowl refuges in the midwest. The well-being of these geese was secure on the Gulf Coast as long as they dispersed themselves over millions of acres of prime wintering grounds. Now they may be herded into small refuges and are threatened with disease, over shooting, high crippling losses, and severe winter losses."

Waterfowl biologist Hugh Bateman wrote that as many as 400,000 blue and snow geese—nearly half the Gulf Coast flock—have been counted on these federal refuges (Sand Lake in South Dakota, De Soto on the Iowa-Nebraska border, and Squaw Creek in Missouri).

Federal officials deny that birds are deliberately held on the refuges for hunting. They point out that they serve a constituency of 50 states—including Louisiana. They tell of diligent attempts to disperse dense concentrations of birds, or shoo more birds southward. They concede that this, if successful, might bring them complaints from states to the north. "We're on the griddle," one told me frankly. "We've got to address ourselves to short-stopping."

The short-stopping controversy flares when an epizootic—the animal equivalent of an epidemic—breaks out on a refuge that concentrates large numbers of birds. If not checked, an epizootic could virtually wipe out a flock. On January 7, 1964, about 20,000 snow and blue geese returned to Squaw Creek National Wildlife Refuge from an early evening feeding flight. The next morning, 1,110 dead geese were found, each still sitting upright, so *(Continued on page 78)*

Managed but menaced: Hunting regulations adequately reflect current abundance or scarcity of these ducks, but their habitats undergo continued disruption by logging, water pollution, and wetlands drainage.

Like waterfowl, deer in the United States exist as public wards, carefully controlled and cultivated for the sportsman's pleasure. The cycle of deer management, symbolized here by New Jersey's prosperous program, begins in the foreground with adult white-tailed deer. Their fall mating produces a fawn in late spring or early summer. Responsible for maintaining a constant deer population in a steadily decreasing area, state biologists and game managers discuss how best to provide a favorable environment for their new charge. Behind them an older fawn—lured into a restraining box with apples and corn—receives an ear tag. Field researchers will note its condition and weight before releasing it. Data on numbers, range, habits, and health of the herd create a basis for sound management. Over the summer, the deer mature, foraging through hemlock, rhododendron, and cedar swamps,

PAINTING BY ARTHUR LIDOV

farmland and pine barrens. In autumn, hunters come to prune the herd that they—through taxes and hunting-license fees—have financed. The gutted carcasses of their fresh kills hang on the scaffolding behind them. Ironically, sportsmen's demands kept the whitetail from becoming a zoo specimen. Their powerful associations began pushing stringent hunting laws through state legislatures before 1900, easing pressure on plundered herds. Often game managers now must cope with the opposite problem: too many deer. In New Jersey, hungry deer overbrowse their range, ravaging prized Atlantic white cedars and other trees. If not killed by hunters, many starve in winter. Although resurgent deer herds have prompted a reappearance of cougars to prey on them in the east, cougars and wolves decline elsewhere in the United States, leaving man a deer's number one enemy—and friend.

Only archers may hunt the white-tailed deer in Cape Romain National Wildlife Refuge, and there only on Bulls Island, a barrier isle for the South Carolina coast. From the crotch of a live oak, an archer takes aim through vines and Spanish moss and draws his bow. Camouflaged hunters carry a doe they bagged as it ventured toward the dunes.

Cape Romain staff keep close watch on the Bulls Island herd by maintaining browse study areas; stripped forage would indicate an overabundance of deer. They issue bow-and-arrow permits without bag limits, as state regulations allow. The refuge officials banned guns in deference to public opinion, especially public fear of accidental injury to a nesting bald eagle.

In Utah's plateau country, inquisitive mule deer—two does and a buck—eye an intruder on their snow-sprinkled domain. Mule deer still roam the West where Indian hunters of the past invoked them in chants and song appropriate to the year's changing seasons.

RONALD A. HELSTROM (BOTH, ABOVE) AND PERRY SHANKLE, JR.

swift were their deaths. All apparently died of fowl cholera. "The continued hoarding of huge concentrations of blue and snow geese on small refuges late into the winter," wrote Hugh Bateman, "would seem to be just asking for similar and perhaps even more severe disease outbreaks."

Early in January 1973, mallards began dying at the Lake Andes National Wildlife Refuge in southeastern South Dakota. During one week about 1,000 mallards were dying each day. Of 100,000 mallards in the area, some 40,000 perished. As the survivors began migrating north in March, biologists expressed fears that they included carriers of the disease, duck virus enteritis. That might mean new outbreaks the following winter.

The disease, also known as Dutch duck plague, first appeared in the United States in 1967, when it struck the Pekin duck industry on Long Island. Never before had it been known to sweep through a wild flock. But, because the Department of Agriculture had jurisdiction over "exotic" (imported) diseases, the mallards of Lake Andes became wards of two federal entities: Agriculture, and the Department of the Interior's Bureau of Sport Fisheries and Wildlife, which runs the refuge.

I no longer wonder whether waterfowl live in a natural or a bio-political world.

The duck stamp, which helped create that world, has delivered more than real estate. (By July 1971, land acquired by duck-stamp funds amounted to 977,000 acres of federally-owned waterfowl habitat and 832,000 acres of private lands opened to hunting.) It has also made waterfowl hunting a powerful factor in determining priorities of wildlife management. Conservation agencies realize the economic reality: more than 2,000,000 stamp-buyers, multiplied by the current $5 price for a stamp, plus the average $150 a hunter spends annually on gear, plus the license revenue of the state in which the hunter seeks his ducks and geese.

The financial contributions of the duck hunter are obvious; those of the non-hunter are not. Taxes, of course, support the Bureau of Sport Fisheries and Wildlife, but the high visibility of hunter revenue is mirrored in the Bureau's budget. The Fish and Wildlife Service focuses most of its research on waterfowl and doves. And many Bureau-managed refuges allow hunting.

The Horicon National Wildlife Refuge in southern Wisconsin, for example, is as much a gathering place for hunters as it is for Canada geese. In 1972, within the officially designated "Horicon Zone" around the refuge, the Canada geese kill quota was 16,000 birds in 18 days. It takes a taste for irony to savor fully the official signs posted at federal wildlife refuges. The signs show a Canada goose—in flight.

Dr. Glen Sherwood, a biologist formerly with the Bureau of Sport Fisheries and Wildlife, writes that our major goose refuges "have one thing in common: too much success in attracting and holding geese." In 1944, he recalls, only 250 migrating blue and snow geese stopped at the Sand Lake National Wildlife Refuge in South Dakota. By 1969, the number had grown to 160,000,

FRANK R. MARTIN

pronghorn

and in that year's season he estimates that 50,000 geese were killed by hunters at the refuge boundary.

Waterfowl managers often cite the comeback of the Canadas as a success story. But Dr. Sherwood's *too much success* haunts the saga's modern chapters. Critics are questioning our stewardship over waterfowl: Are we saving birds merely to kill them? Are our refuges actually game-bird factories surrounded by shooting galleries? Where is the ethical boundary between harvest and slaughter?

Most managers I talked to shrugged off such questions as coming from emotional, and usually ignorant, non-hunters—"the thou-shalt-not-kill crowd," one federal official called them. But not all the critics are naive. Dr. Sherwood is not. Neither is Dr. H. Albert Hochbaum, author of the definitive *Travels and Traditions of Waterfowl*. In 1970 he called upon the government to return to its "major function as protector rather than provider of waterfowl."

He was cited by Dr. Sherwood in a report on a new "providing" by game managers: previously non-hunted species. The 1916 agreement with Canada had prohibited for ten years the hunting of certain rare birds, including cranes and swans.

The populations of some built up through the decades until New Mexico, proclaiming such success with the little brown crane, requested permission to open a season on it. In 1961 the little brown crane lost its immunity in the skies of New Mexico. Now it is also hunted in Texas, Oklahoma, Colorado, North Dakota, South Dakota, Montana, Wyoming, and Alaska. Since 1964 Canadians have been permitted to bag little brown cranes in Saskatchewan and Manitoba.

Bear River Migratory Bird Refuge in Utah attracts one of the continent's largest concentrations of whistling swans. Not surprisingly, then, it was in Utah, in 1962, where thou-shalt-not-kill protection ran out for the whistling swan. In that first season the quota was 1,000. In 1969 the quota rose to 2,500, and hunting was authorized in Nevada (500). Most of Utah's swans fell to hunters on a state refuge, but many were hunted where they had found sanctuary since 1929—at Bear River.

On that chill day when I stood at Mervin Cohey's blind, a wedge of whistlers sliced through the low-hanging fog above us. *Invulnerable,* I wrote in my notebook, for they seemed to fly boldly, as if somehow they sensed that they could not be pulled from the sky. I did not know then that some hunters and farmers in Maryland had been campaigning to make the whistling swan a target for the guns.

I know now that no animal is invulnerable. Some have become so dependent upon man that their species will vanish if they do not submit to human plans. For them, independence means almost certain extinction.

Newborn pronghorns rest with their mother at a Montana refuge. Hunting regulations help safeguard pronghorns today. But the officially endangered Sonoran race, with a few in Arizona and perhaps a thousand in Mexico, needs international cooperation for its survival.

4. "We're Going to Lose Some...."

Recovery plans bring federal resources to the intricate tasks of saving the most endangered

Jim Shaw tilted his head to the starry Texas sky and let out a howl. A chill rippled along my spine as the howl faded into the silence of the night. With Jim and photographer Ron Helstrom, I stood in the middle of the lonely country road, listening. I could hear only the wind softly strumming invisible telephone wires. Jim gulped a swig of ginger ale—his remedy for a throat raw from howling—and tried again. Silence. But maybe . . . I thought I heard it. So faint, so far away, I could not trust my ears. Maybe. We all looked into the blackness, as if eyes would help us hear. Finally, Jim shook his head. "Not good enough," he said. "They just aren't responding tonight."

In our car, driving to the little town of Anahuac, Texas, Jim smiled and said, "Same thing happened when I had my doctoral adviser down here. They wouldn't howl for him, either." Jim had been howling for hours, trying to evoke a response from a phantom: the red wolf, one of the rarest mammals in North America. Tired and hoarse, he settled back in the seat. It had been a long and wolfless day. But he knew they were there in the dark sea of grass that stretched for miles around us.

On other nights he had drawn answers—long, steady howls

With a shrill, yapping howl, a red wolf-coyote hybrid answers her mate. While predator control eliminated the red wolf over most of its range, a crossbreed population developed—notably in central Texas.

C. C. LOCKWOOD

that hung for as long as 16 seconds in the air. Sometimes the answering howls grew louder as the wolves cautiously approached, and sometimes the responses changed to sharp barks of alarm. If any animal knows man to be an enemy, that animal is the wolf.

Ever since the white man arrived on the continent he has been at war with the wolf. But here, along a patch of Texas coast, a kind of truce is being negotiated. The truce is a phase of the Office of Endangered Species' first recovery plan, an unprecedented attempt to bring back a species rather than merely to save a few of its members. The plan is the prototype of others to come, each designed to restore an imperiled animal.

Given the task of saving more than 100 species, the Office must establish priorities. "We're going to lose some," one official conceded off the record, citing the scope of environmental change that threatens them. "But there are some that we should not lose at any cost."

I had gone to Texas to watch a species recovery team in action and to view the unfurling of the plan, which involves the State of Texas, biologists like Jim Shaw, ranchers of varying allegiance to the wolf—and a federal agency whose usual mission is the *killing* of wolves.

No one who hears the drum roll of problems can claim that O.E.S. has chosen an easy one for its first recovery plan. The task, as described by its architects, sounds herculean: "We start with a nucleus of about 300 animals that are physically weakened by intestinal parasites, heartworms, and mange. They cling precariously to [2,000 square miles] of coastal prairie—much of which is too wet for year-round habitation. Virtually all of the red wolf range is in private ownership and there is widespread public apathy. . . ."

With Jim, a graduate student at Yale, I set out to see some red wolves on the Anahuac National Wildlife Refuge. This heart of wolf country is on East Bay in Chambers County, about 50 miles southeast of Houston. Walking down shell roads—named for the oyster shells that covered them—and wandering the marshlands, we did see "wolf sign": tracks pressed in the mud, and scats, dung with bristles of fur that revealed a wolf's meal of nutria. This muskrat-like animal, a South American import introduced as a possible fur resource, has helped save the red wolf. As long as it can get nutria and rabbits, say old-timers around here, the wolf is likely to stay away from calves.

We saw Canada and snow geese rise, honking, to the sky. We saw roseate spoonbills in a grove of salt cedars. We saw nutria plunking into canals along the road. We saw a skunk lazily ambling along. But we saw nothing resembling a wolf until we walked up to a manmade hummock that is the highest, driest bit of land in the refuge's 9,836 acres. On it stood a line of cages. And in the cages: red wolves? Once again, maybe.

Three of the seven captives were coyotes, kin of wolves. Pointing to a larger animal, Jim said, "Big. But probably a hybrid." Another, a bit smaller, cowered in a box in his cage.

AGNES STOUFFER

MARTY STOUFFER

RUSS CARMACK

Four days after whelping, the hybrid shown on page 80 carries her young from one den to another in a half-acre lot enclosed for filming. Photographer Marty Stouffer holds month-old pups he identifies as a coyote in his right hand, a hybrid in his left. Experts unfamiliar with them and their parents say they cannot give an identification. But all agree that the adult male (left) captured in Chambers County, Texas—key area in the Office of Endangered Species' red wolf recovery plan—has the long legs, big ears, and large body of the pure red wolf.

"That's Pegleg," Jim said. "O.K. But he's scrawny." (Pegleg had lost part of his right forefoot in a trap.) The other two? "Maybe."

How do you know for sure? Clark Bloom, of the refuge staff, shrugged. "We get Glynn Riley to check that. I've been here three years and I don't think I've seen a dozen." Distinguishing red wolves from hybrids takes experience.

In the days that followed I would hear much about Glynn Riley, a legendary trapper who seemed to have a sixth sense about the red wolf. He could, I was told, "read a scent post," a bush or patch of ground or clump of weeds on which wolves urinate, in effect leaving calling cards that record their passage through an area. Glynn sniffs it to learn whether a wolf has marked it within 12 hours or so.

Even before I met Glynn, I decided that reading a scent post might be easier than recognizing a red wolf. Its identity as a species has been clouded by hybridizing: cross-breeding with coyotes, and coyote-like canids.

To save the red wolf, its rescuers had to define it biologically. Then, according to the recovery plan, the true red wolf must be kept from mating with other canids. Unless man intervenes, the red wolf may breed itself out of existence.

Biologists reached this conclusion by a circuitous route that goes back in time to the 19th century, winds along dusty shelves in the Smithsonian Institution, passes through a BMD07M computer program, and continues down country roads where Jim Shaw howls, to refuge cages where mysterious animals await Glynn Riley's verdict.

In 1915, when the Federal Government first joined stockmen in their war on the wolf and coyote, scientists at the Smithsonian asked for the skulls of the animals. These were added to earlier collections that taxonomists could use to sort out the wild canid species of North America. Most authorities recognized three: the coyote, *Canis latrans;* the gray wolf, *C. lupus;* and the red wolf, *C. rufus*. Taxonomists, classifiers of living things, use a binomial system of Latin names that incorporates the genus and the species. If necessary, they designate a subspecies with a third Latin name. A species is a group of animals that usually breeds only within itself; a subspecies or race generally arises as a geographical variant.

In 1905, a biologist had reported a small red canid from south-central Texas. Was this really *C. rufus?* Or merely a subspecies of the much bigger gray wolf that roamed the plains to the west? Or a subspecies of coyote?

Such questions made little difference to federal trappers, who began their eradication of four-footed varmints in eastern Texas in 1915, in Oklahoma and Arkansas in 1918, and in Missouri in 1922. They labeled skulls with dates and locations and shipped them to the Smithsonian, which duly stored them.

Not until the early 1960's did many people worry about the disappearance of the red wolf. Even then the Bureau of Sport Fisheries and Wildlife, which administered the predator-control program, was reporting the killing of thousands. The

PATRICIA CAULFIELD

Federal trapper and friend of the red wolf, Glynn Riley cleans a newly captured—and fully tranquilized—hybrid male. Years of experience in the field enable Mr. Riley to identify a true red wolf at a glance.

supply seemed inexhaustible. In 1962 a biologist speculated that the wolf was near extinction in Texas; the next year the Bureau reported a red-wolf kill of 2,771. But as the concern about extinction grew, the Bureau decided to investigate.

John L. Paradiso, a Bureau biologist working at the Smithsonian, got the assignment. Skulls of "red wolves" killed in east Texas in recent years were sent to him at the museum. He learned that the Bureau had arbitrarily set a boundary—the 100th meridian—as the dividing line between red wolves and coyotes. Any animal killed west of it was called a coyote; any to the east, a red wolf. The 100th meridian bisects central Texas and runs the length of the Edwards Plateau, sheep country and a prime wolf-killing zone.

"It soon became obvious that these were not the skulls of red wolves," John Paradiso told me. We were in the Office of Endangered Species, to which he had recently transferred. His narrative reminded me of my police reporter days when I listened to a good detective recount the cracking of a tough case. He always made it sound easy.

"These were not robust, wolf-like animals at all. You couldn't really say they were red wolves, and yet they were too darn big to be coyotes.... You'd get a little tiny one which is identical to a coyote and you'd get a great big one which looks just like the old red wolf, and you'd get all these in-betweens. The way we interpreted it, the red wolf and the coyote were hybridizing.

"Around the turn of the century there was extensive habitat modification—and in the Edwards Plateau there was a big campaign to kill wolves. So those two things combined. There apparently just weren't enough red wolves for each to find a mate."

I asked how he pictured what happened. "We think a small female red wolf, not finding a mate of her own species, would mate with a large male coyote. These animals are so closely related that the young were perfectly fertile and they would mate with coyotes too. And so a hybrid swarm developed on the Edwards Plateau—and elsewhere. Meantime, most of the red wolves had been killed out of their former range. The growing hybrid swarm moved in and filled the canid niche that was formerly occupied by the red wolf.

"Now what you have in most of east Texas and going right on into Louisiana and Arkansas is this hybrid swarm. All carry the genes of the coyote and the red wolf."

Usually, he explained, hybrids do not long survive as a population; they do not fit into the niche of either parent. But, as the hybridizing increased, so did the disruption of the environment. Inadvertently, man was making new and unnatural habitats in which these new animals could exist. The hybrid found many homes. It became more and more coyote-like as the red-wolf killing campaign eliminated potential mates of that species.

One of John Paradiso's colleagues, Dr. Ronald M. Nowak,

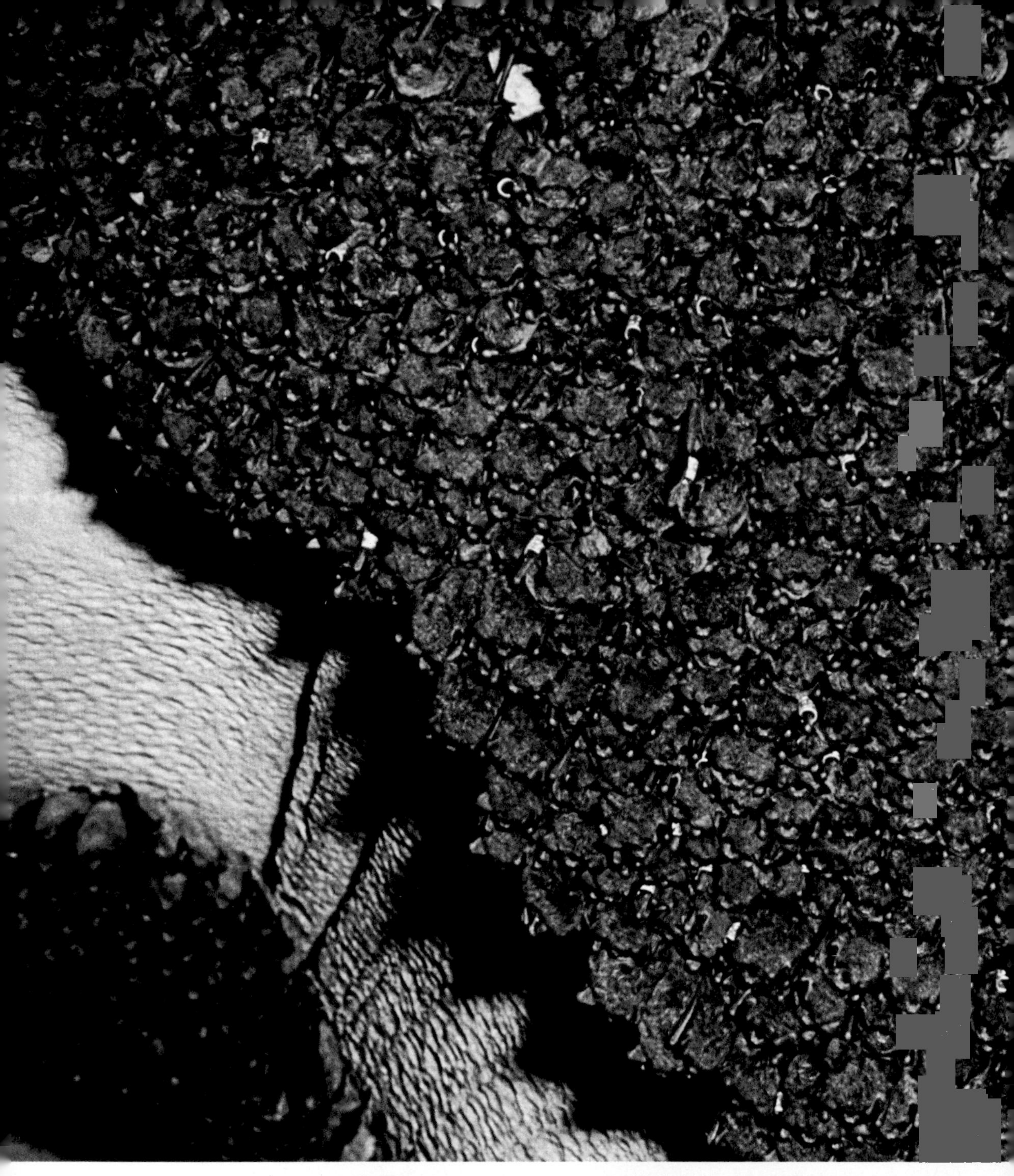

Indiana bat

spotted bat

ROGER W. BARBOUR

Virginia big-eared bat

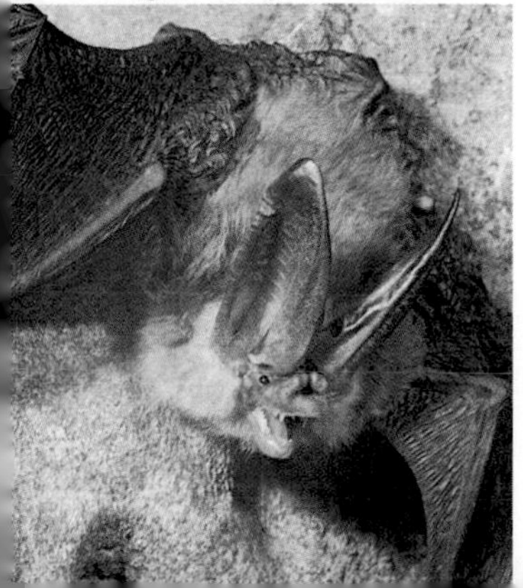

Hanging head down until spring, a thousand Indiana bats cling to the ceiling of Bat Cave in Kentucky's Carter Caves State Park. Of 500,000 bats in this population, most winter in just three caves and one mine. An iron gate now defends Bat Cave's denizens from wanton vandalism. Sheer rarity gives the southwest's spotted bat its claim to attention; sheer ignorance of its life hinders measures to preserve it. The Virginia big-eared bat haunts Appalachian caves—but abandons those disturbed by human intruders. Without protection of their habitat, this entire race could perish.

then a University of Kansas graduate student, turned to a computer to trace the decline of the red wolf. With a chronological series of hundreds of Smithsonian skulls, he used a computer program: 15 measurements from each of the skulls, along with standard skull measurements for the coyote and the red wolf. The computer gave back to him a picture that showed which animals were coyotes, which were red wolves, which were hybrids. Then, with maps, he turned to data from the labels: where each animal had been killed, decade by decade.

Plotted on Ron Nowak's maps, the shrinking range of the red wolf conjures the image of a once-great army slowly retreating and collapsing before the relentless forces of an enemy. In 1800 the outer line of the red wolf's range swept through eastern Texas and Oklahoma, arching across Missouri, Illinois, Indiana, and probably eastward through the Carolinas. By 1900 the perimeter encompassed only a swath of southern states. By 1960 it ran less than 50 miles from the shore of the Gulf; there the remnants of the species made their last stand.

The first solid evidence that somewhere a pure red wolf population still existed came in 1964, when the Smithsonian received a few new skulls—"enormous, just like the old red wolf skulls in the collection," John Paradiso recalls. They came from around the Anahuac refuge. He went to investigate.

A relatively desolate region, sparsely settled, the Anahuac area was a pocket where red wolves could survive. "All signs showed that the animals there were all big," John told me. "There were no in-betweens, no hybrids. We concluded that in that area, and perhaps going into Louisiana, there was a pure population of red wolves."

But now the hybrids are invading this region too. Where John Paradiso had found a pure population there are reports of crossbreeds—"strange-looking canids and smaller animals." Many of the reports have come from a trapper named Glynn Riley.

I finally met Glynn at his home in Liberty, about 20 miles north of Anahuac. As we spoke, I could feel the soft, warm touch of the wolf. A pelt, gift of a local rancher, was draped over the sofa I sat on. Wolf skulls were arrayed on tables and shelves around the room. Glynn Riley, in a way he finds hard to describe, loves the wolf. He was a trapper by trade—now, officially, a biological technician for the Bureau's Division of Wildlife Services. The division deals with predators for ranchers and farmers.

Glynn has trapped wolves, coyotes, and other predators for a living ever since he was a teen-ager. "I don't know any of us who wants to see the animals die, who enjoys it," he said. "The Good Lord made them predators and when you put temptation in an animal's way, you can't blame him for what he does. But if he's taking money out of somebody's pocket, then you can't blame that man for wanting that predator out of there." It was a long speech, for Glynn Riley.

When I asked him how he tells a red wolf from a hybrid, he said, "If you've got to scratch your head, it's no red wolf. It's a hybrid. Sometimes there's something out of proportion, head or body or legs too large or too small. Usually, there's just something about him that's odd."

I could understand why the recovery plan's passage on identification sounds so politely exasperated: "The problem is that what appears to be a true red wolf to one observer is assumed to be a hybrid by another. To date, only the personal judgment of highly competent field personnel has been used to separate hybrids from 'true red wolves.' "

Jim Shaw points out that these competent personnel consistently agree on their identifications. And maybe his howling technique will help determine how many wolves there are. He estimated, by howl-counting, that there are about 60 in Chambers County alone. (Glynn says, "There's more than people think. Only the Good Lord knows for sure, but I'd guess there's about 300.") Or maybe blood analysis or sophisticated genetic tests will clarify questions about hybrids.

The *maybes* go on. Meanwhile, the plan focuses on other problems. "We have to go in and livetrap all the canids in a good-sized zone," John Paradiso says, "so that the hybrids don't flow in." If a trapped animal is judged a red wolf, it will be released in the heart of the large "management area" the plan calls for. If it is not a red wolf, its fate is still uncertain.

Local attitudes will have their bearing. One lifelong resident, rancher Joe Lagow, favors conservation. When I called on him one day with Jim Shaw and Ron Helstrom, he picked luscious tangerines for us, and introduced us to his cattle, his pet deer, his collection of exotic game birds, and the alligator that lives in one of his ponds. He also told wolf stories.

"Right here in back of the house we have a 50-acre improved clover field. My father-in-law and I were goose hunting there one day. A bunch of 15 or so Canada geese lit just out of range. Well, here come a couple of wolves. One jumped the fence and the other crawled. The old bitch wolf, she got down in the grass. The old dog wolf walks around to the west. He just trots along, and he drives those geese right near where the old bitch was. Then he makes a charge. Well, they took off—right over the clover—and she jumped up and got a goose that was *flying*, flying right over her, and ran out of the field.

"I want to protect the wolf," he said. "I try to protect any wildlife, as long as it doesn't try to destroy me."

Converting local wolf-haters into wolf-lovers sometimes takes a bit of psychology. Once, when Glynn Riley put out a line of traps, a few property owners volunteered to watch them for him. When he came back, though, he found a wolf dead in one of the traps. It had been shot three times. Glynn found the men responsible. But, instead of berating them, he passed out catch poles and transfer cages. He convinced them that getting a big live wolf out of a trap and into a cage called for more skill —and sport—than shooting it in a trap. Later they accepted the challenge, and, as Glynn says, "Now they have a wolf story that none of their rifle-toting buddies can top."

The standard trap now in use has offset jaws, to prevent

Carefully preening, a whooping crane keeps its dazzling feathers fit for flight—although minor wing surgery has confined this majestic bird to the Patuxent Wildlife

PATRICIA CAULFIELD (ABOVE) AND NATIONAL GEOGRAPHIC PHOTOGRAPHER OTIS IMBODEN

Research Center in Maryland. The fight to save the whooping crane from extinction grew desperate in 1945, with 17 counted at their winter home, the Aransas National Wildlife Refuge on the Gulf Coast. A seven-foot wingspan carries the snowy whoopers (above); accompanied by their gray cousins, sandhill cranes, they rise from a feeding area at Aransas. In spring the few dozen whooping cranes migrate 2,500 miles to breeding grounds in northern Canada. The flock can feed and rest at refuges along the route, but dangers remain—such as collision with power lines. And in autumn, mistaken identification by hunters poses a threat to these magnificent cranes.

painful wounds, and padded jaws are being tried out. A local veterinarian, Dr. Aaron Long, treats any injured or sick red wolves as a public service. He is to get reimbursement for expenses, but no fee; the plan has no funds for it.

"There's plenty of money for waterfowl and deer," one of the wolf's frustrated champions complained to me. "Game animals never seem to be in need. But wolves don't have feathers or horns. And they have long fangs."

The Office of Endangered Species originally estimated the cost of the three-year recovery plan as $99,200; it will almost certainly go higher.

I first saw a draft of this plan taped on a wall in the Office of Endangered Species. It stretched across a paper seven and a half feet long and contained 190 steps. Pondering it, I was overwhelmed by the complexity of saving an endangered species. And, standing before the cages on the Anahuac refuge, I realized how far it is from a plan on a wall in general headquarters to an animal on the front lines.

I have seen other recovery plans, each with numerous steps, each step describing in a few words an action that may be incredibly difficult or expensive to perform. To believe that some of these plans can be carried out, you need the kind of faith that moves mountains—or cleans up Lake Erie. That is one of the "steps" needed for the restoration of the blue pike.

Gene Ruhr of O.E.S., a man of great faith, says cleaning up Lake Michigan would be *really* difficult because it is so much deeper (275 feet average) than Erie (60 feet average). "For one reason, Erie's a shallow lake with a fairly high flow-through," he explains. "Once we cut off the sources of pollution, there's a much better chance that the lake will come back. So, if we can find the blue pike and get it started somewhere else, then we'll wait for the lake to clean up."

Like the red wolf, the blue pike raises questions of identity—it interbreeds with the closely related walleye. There were certainly "good," or pure, blue pike as recently as 1955, when 19,700,000 pounds of them were hauled out of Lake Erie. With no restrictions on the catch, and with pollution increasing, the lake's most important commercial fish began disappearing, rapidly. Though fishermen kept reporting catches of blue pike, biologists identified them as small walleye.

PERRY SHANKLE, JR.

In 1969, what were believed to be a pair of good blue pike were successfully bred, and some 9,000 fry were shipped to a hatchery in South Dakota. An unsuccessful attempt was made to stock a lake in Minnesota to build a population for the time when conditions in Lake Erie improved. With this failure, and with a question about the purity of the parent stock, the plan is left searching for good pike—and fit habitat.

Some recovery plans are less awesome. The Indiana bat, for example, numbers about 500,000; the majority of them hibernate in just three caves and one mine. A few years ago, two boys wandered into one of Kentucky's Carter Caves and beat 10,000 of these bats to death with sticks. The state helps protect this endangered species with an iron gate and chain-link fence

Subject of exhaustive efforts to save its species, a wild whooping crane trumpets a wail at Aransas. Research at Patuxent attempts to establish a breeding group of captives, insurance against a disaster in the wild. Sandhill cranes, close kin of the rare whoopers, act as stand-ins for experiments. Kathleen Kennedy hand-feeds sandhill chicks in a wire ring that isolates them from greedy turkey poults. The cranes start fighting as soon as they can stand; the poults divert them from attacking each other. Preparing a hundred pounds of feed for young cranes, Dr. John A. Serafin, a research nutritionist, adds soybean meal to other ingredients needed for a balanced diet. Handlers shroud themselves to keep incubated cranes from mistaken imprinting —identifying the first moving object they see as a parent—and possible failure to mate as adults.

N.G.S. PHOTOGRAPHER OTIS IMBODEN (BELOW AND UPPER RIGHT)

LUTHER GOLDMAN, BUREAU OF SPORT FISHERIES AND WILDLIFE

at the cave entrance. Now about 50,000 Indiana bats winter there in relative safety. Nature, incidentally, wrote the recovery plan for another bat, which may be the rarest mammal in the United States. The Ozark big-eared bat—there are believed to be no more than 100 of them—stays alive by staying scarce. "So long as it remains unknown to man," says the Office of Endangered Species, "this race will probably persist."

Some animals will indeed survive because they remain unknown. But most now need human help, usually to counteract human assaults or neglect in the past.

Long before the red wolf recovery plan was unveiled, the National Audubon Society, the U. S. Fish and Wildlife Service, and various Canadian authorities undertook to save a species that was plunging toward oblivion. The "Cooperative Whooping Crane Project" was launched in 1945. An incomplete count for that year stood at 17, the number that had reached their winter home, the Aransas National Wildlife Refuge on the Texas Gulf Coast, after a 2,500-mile flight from their breeding ground in northern Canada.

Twenty-eight years later, the whooping crane was still in trouble. In 1972, of 56 that went north, only 46 returned to Aransas—a loss of ten adults from the year before. Few if any other animals had been more closely watched than this great white bird. Yet no one knows what happened to the lost ten. Not even the Royal Canadian Mounted Police. They tracked down one reported sighting; it was the body of a white pelican.

Nowhere is the loss felt more deeply than with the Endangered Wildlife Research Program at Laurel, Maryland, where the whooping crane reigns as the supreme imperiled species. Seventeen live there in regal protection, more precious now than ever. For each loss in the wild emphasizes the captives' designated role: preservers of the species. The program, part of the Bureau of Sport Fisheries and Wildlife's Patuxent Wildlife Research Center, functions as a gene bank for endangered species. Whooping cranes are deposited there as eggs. Raised to produce stock for release to the wild, they are investments in the future.

Dr. Ray C. Erickson, head of this research program, took me out to see the whooping crane area, but only from a distance. The birds strutted in enclosures on a hill about 100 yards away. I had to be satisfied with close-up views of their stand-ins, sandhill cranes that flared their wings and croaked in alarm—*garooo-a-a-a* from the males or *tuk-tuk-tuk* from the females—when we approached their pens.

The sandhills, cousins of the whoopers, are what Ray Erickson calls surrogates. In 1961, he began experimenting with the sandhills, a much more plentiful species, expecting to apply what he learned to the whoopers.

"We tried to trap young sandhills on the wintering grounds," he told me. "But some birds were killed by the nets. We couldn't take such risks with the whooping crane." The experimenters next tried taking chicks. Most of them were suffering from a respiratory disease caused by a fungus. Egg-snatching, the one method of capture left, turned out to be the safest.

After six years of research the Erickson team had learned enough about taking, transporting, and incubating sandhill eggs to try their technique on whooper eggs. In May 1967, American and Canadian biologists collected six eggs from the whooping crane's only known nesting ground, Canada's Wood Buffalo National Park. Named for the wood bison, another endangered species, the park covers 17,300 square miles—Connecticut, Massachusetts, Rhode Island, and Delaware could easily fit inside it. A breeding pair of whooping cranes seeks isolation, but its rescuers knew where to find the nests in that vast wilderness 400 miles below the Arctic Circle.

Five eggs survived the flight and were hatched at Patuxent. The hatchlings were borne from incubator to pens, and tended, by men who looked like Halloween pranksters. The chick-bearers had draped themselves in white sheets to blur the whoopers' image of man. Animal behaviorists have learned that the young of many species "imprint" on the first object they see, often identifying it as a parent. The staff feared that their sandhills might imprint so strongly on the keepers that as mature birds they would ignore each other and never mate. The Patuxent whoopers were shown nodding, bowing wooden silhouettes of their "parents" in the hope that they would imprint on this image—and not suffer an identity crisis as adults.

Successful hatching of those five eggs at Patuxent meant that only half of Ray Erickson's worries were over. What if the eggs left in the nests were deserted? The answer came in the fall of 1967 when nine young whoopers, a near-record complement, arrived with the migration flock at Aransas. Dr. Erickson now knew that one of his carefully tested theories had proved valid: Late in the incubation period, whooper parents—like sandhill parents—will not desert the egg left behind in a raided nest.

Egg-taking had no apparent effect on the hatching and rearing of chicks at the breeding grounds. In fact, Dr. Erickson told me, parents usually raise only one of their two young, anyway. "They have a good hatching rate, but something happens between hatching and flight. There are too many young birds lost in the wild. Our record is a 90 percent hatching success—which is about as good as the old lady herself can do."

Nobody knows whether the "old lady" has the problem of chicks' fighting and killing each other. Captive chicks begin fighting as soon as they can stand, but in Patuxent pens they don't have to fight each other. Turkey chicks have been introduced to divert a whooper from fighting a valuable relative. Turkeys can outrun cranes, so after a while the fighting eases; and all the chicks benefit from the exercise.

The feeding and care of whoopers has become such a science at Patuxent that Ray Erickson believes his captives are likely to live much longer than wild whoopers. He also has faith that his birds will be more productive than wild ones.

Probably the whooping crane was never really abundant. It seems conspicuously unable to adapt when change—natural

Flaring full-curl horns—prized hunter's trophy—crown this desert bighorn ram. As civilization claims the domain of desert sheep, and as feral burros and livestock compete for space and water, conservationists struggle to halt their decline. Four transplants in process—one a transfer from Mexico—have so far proved successful. On some federal lands the Park Service maintains artificial watering stations. Friends of the desert bighorn hope such measures will help save it from the fate of a Badlands subspecies named for Audubon—first bighorn seen by the Lewis and Clark expeditionists in 1805—which disappeared around 1910.

C. C. LOCKWOOD, NATIONAL AUDUBON SOCIETY

or man-made—affects its preferred habitat. Now, with its southern home so small, its wild population increasing so slowly, the captive flock at Patuxent is genetic insurance against a catastrophe anywhere in 2,500 miles.

The stated objective of the Patuxent propagation program is to "establish a captive breeding group of whooping cranes with which to produce a stock for the reestablishment of breeding populations in the wild."

Dr. Erickson thinks the objective is almost within reach. Sandhills, again performing as stand-ins for whoopers, are now laying at the rate of six eggs per pair per year. Incubation hatches two out of six successfully. Experiment after experiment—artificial insemination, "lengthening" breeding days by floodlights, changes in diet—finally they are paying off. Initial disappointment now yields to cautious optimism.

Still, in captivity a bird faces not only many of the survival problems it would encounter in the wild but also a host of others: stresses produced by experiments, by handling, and by being moved about; diets that prove inadequate; diseases that are penned in with the captives; injuries suffered in escape attempts. The raising of wild animals, so often frustrating and heartbreaking, is also controversial.

The prospect of continued removal of whoopers' eggs from nests in the wild has its critics, including the National Audubon Society. In general, "protectionists" argue for environmental protection of wild species rather than "propagationist" strategy of "take and put"—take eggs out, put birds in.

But Dr. Erickson doesn't see it that way, and his researchers are experimenting with releasing captives to the wild. The Florida sandhill crane, one of those 69 creatures in a kind of federal limbo, and the Aleutian Canada goose, an endangered subspecies, have been matriculated into wildlife society.

The return of the Aleutian goose recalls the myth of the phoenix arising from its own ashes. The small relative of the Canada goose bred in safety on many of the Aleutian Islands until foxes were introduced for their fur. The foxes ravaged the ground-nesting geese, which had never known such predators there. One small colony, on tiny Buldir Island, escaped because their breeding site was far out on the Aleutian chain.

In 1963, men invaded Buldir, but these were saviors—Fish and Wildlife Service biologists. They easily captured 16 goslings. The geese were reared and bred at Patuxent to become the progenitors of new colonies on their old breeding islands.

The Patuxent restorers kept their geese flightless by clipping their wing feathers. Then, early in 1971, the feather stubs were pulled out to stimulate regrowth. Seventy-five geese would fly—first by plane to Amchitka Island, which had been cleared of foxes—and then, on their own wings.

As the freed geese began foraging in the tundra for food, a few bald eagles appeared in the sky. The eagles killed four geese around the release area. The others, quickly growing wary, began flying farther away and soon vanished. With them soared the hopes of the staff at Patuxent who had given them wings.

On their way to their wintering grounds, Rocky Mountain bighorns pause on a ledge in Glacier National Park, Montana. Swiftcurrent Lake lies 1,500 feet below. The sheep may descend as far as the lake to find winter forage; in summer they return to the peaks. Outside the park, the opening of public rangeland to livestock has jeopardized other herds. Bighorns prepare to mate soon after fall migration. At right, a ram performs a courtship ritual called lipcurl or "flehmen" to determine the receptivity of the ewe beside him. After smelling her urine, he tilts back his head so the odor can reach a special nasal organ for analysis. If it indicates estrus—the fertile period—mating will begin. The ewe's grazing shows a common reaction of these mammals to stressful situations.

BILL MCRAE (ABOVE); JEN AND DES BARTLETT

To see Patuxent's most closely guarded captives I had to slosh through a shallow tray of disinfectant, wait until Ray unlocked a gate, pass through a fence, and enter a metal shed big enough to hold a small herd of cattle. But the only animals here were in a small wooden box suspended about three feet above the floor, behind a barrier. "Not too close now," Ray said. We stood in that big shed, about ten feet from the small box, until out of a round hole in the box popped a head: black mask across big black eyes, glistening black nose in a bristly white muzzle. The head swiveled around, then disappeared.

I had just had my first look at a black-footed ferret. There was another one in the box. Like its fellows in the wild, though, it stayed out of sight. Bats excluded, there may be rarer mammals in North America, but none as rarely *seen* as the ferret.

Nor is any mammal more of an enigma. For nearly three years, using the surrogate technique so successful for the cranes, Patuxent researchers had studied the European ferret, a close relative of the black-footed. Then, convinced they were ready, they negotiated with state officials in South Dakota for six ferrets. As "resident" animals, they were controlled by the state at that time, regardless of their federal status as endangered.

The animals were trapped in 1971 on ranches where their lives were threatened by imminent poisoning of their habitat: prairie dog towns. Because biologists particularly feared that the ferrets were susceptible to canine distemper, anything that would touch a captive was doused with disinfectant. Although about 150 European ferrets had been vaccinated safely, the guardians of the black-footed ferrets did not want to take any chances. They called in experts from the Department of Agriculture, the National Institutes of Health, and one of the laboratories that produced the vaccine. Only then was the decision made: vaccination for the six captives.

Four—all females—died. After months of study, a pathologist determined that the vaccine had killed them. The two males I saw in the box were still inexplicably healthy, but the propagation plan for restoring the ferret certainly was not.

The National Audubon Society, while commending the Bureau of Sport Fisheries and Wildlife for its fieldwork on the ferret, strongly criticized this particular plan. It felt that because the ferret's life seems so interwoven with the prairie dog's, "to study the ferret in isolation in the laboratory may be more of an exercise in academic research-for-research's sake, than a valid and enlightened effort to preserve an endangered species."

Ray Erickson told me the tragic story of the ferret and, in fact, gave me a copy of the stinging Audubon broadside. He is a scientist, an honest man, and, as he says, not one to want all his eggs in one basket. He suggested that I visit his field biologists and find out what was going on beyond the isolation of the laboratory. He thought I would be interested to see what they were up against. One of the biologists was in South Dakota, and he was looking for ways to save the ferrets there.

BETH CHADWICK (ABOVE)

Catching tenuous toeholds with cloven hoofs, mountain goats methodically negotiate a wall of rock. Inhabitants of bleak cliffs and pinnacles, the goats once seemed secure, but new logging roads bring hunters closer and shooting may deplete accessible herds. Below, Washington game officials bind a nanny for transport to Oregon; the burlap blindfold helps to calm her. Oregon, with no native mountain goats, has twice imported nucleus herds for permanent populations. One transplant apparently failed, but a herd in the Wallowa Mountains now numbers approximately 30.

DAVID HISER

5. The Economics of Extinction

For pests and varmints and innocent bystanders, old enemies and some new allies

As we walked through the prairie dog town, exclamation marks punctuated the grasslands: !!! Popping up !!!! Disappearing !!!! Scurrying from mound to mound: !______! Pinning sharp barks to the air: *Yip!!* Alarm! *Chirp!!* Astonishment! When we stopped, the tempo of the ballet slowed and the mounds became as still as the great buttes that walled the horizon beyond the town.

Enthralled by the staccato choreography of the prairie dogs, I had forgotten that we were not here just to see them. I was visiting their 1,000-acre metropolis, near Badlands National Monument in South Dakota, with Conrad N. Hillman, a field biologist for the Patuxent wildlife research program. Day and night, he searches for an animal best symbolized by a question mark, the black-footed ferret. Where is it? How do you find one? How many are there? How do they live?

Soon these may become as hypothetical as questions about the dodo. Unless we can learn enough about the ferret to save it, the species will almost certainly die, its epitaph a cluster of questions.

Con Hillman has found some answers. He warned me,

A black-footed ferret—the only one seen during a two-week round-the-clock watch by photographer and biologist—gazes from a prairie dog's burrow in South Dakota. Always elusive, it may face extinction now.

PATRICIA CAULFIELD

however, that I probably wouldn't get a glimpse of a ferret. "I saw my first three days after coming out here," he said. "But it was three months before I saw the next one." I met him in March; he had not spotted any since the previous November. For one study he spent 1,025 hours watching; during only 144 hours did he see ferrets—often, only the quarry's green eyes reflecting a spotlight's beam. Then he would wait until dawn, hoping his quarry would still be there. I could appreciate his eagerness and his patience, knowing how rare the ferret has apparently always been and learning how imperiled it is today.

"It is with great pleasure that we introduce this handsome new species," John James Audubon and John Bachman wrote in their great work, *The Quadrupeds of North America*. Thus in 1851 did the black-footed ferret make its zoological debut. Their report was based on a single specimen, an animal found in Wyoming, stuffed with sagebrush in a backwoods effort at taxidermy, and then mislaid for years. Until a second specimen was reported in 1876, skeptics wondered if the ferret portrayed in the book existed beyond its pages.

Scattered sightings down the years roughly established that the handsome new species was found throughout the Great Plains and in the Rockies, as high as 10,050 feet. Its range today is still known only vaguely, though undoubtedly coinciding with that of the prairie dog—in grasslands country.

Its peril stems from its position as innocent bystander at a conflict between men and prairie dogs. The ferret does no harm to ranchers who want prairie dogs exterminated; in fact, it eats the rodents that eat the ranchers' grass. Yet the ferret dies—apologetically described as a "non-target species"—in the poisoning of prairie dog towns. How many have died in the conflict, no one knows, nor does anyone know how many there were before the poisoning began or how many now live in towns that will be poisoned in the future.

So Con Hillman's observations are the quintessence of rarity, and reading them I felt the poignancy of seeing an animal that may die off as furtively as it has lived: "Young often played above ground, running in and out of burrows in pursuit of one another. They bit and pulled at each other, humped their backs, and ran on their toes. They scratched themselves frequently, often turning in circles attempting to bite their tail. Often they were seen chasing flies. . . . When approached by the mother, they playfully jumped and bit at her."

Although I never saw a wild ferret, and although Con is a modest, taciturn man, I did learn a little about how to look for ferrets as we toured the towns. Con pointed to a mound whose opening was packed with dirt: "That's a plug. The prairie dog fills up burrows that ferrets have used. Maybe it's a defense against them. I've seen a ferret sticking his head out of a hole and then a prairie dog would rush over and plug the hole." The prairie dog kicks dirt into the hole with its hind feet, then packs down the dirt with its nose.

We crouched over what looked to me like a worn-down mound. Con explained what we *(Continued on page 110)*

Crossing a prairie dog town in South Dakota, a buffalo avoids holes where a misstep could break a leg. Ranchers, fearing loss of grass, want such towns poisoned. But poison spread for the rodents claims—directly or not—"non-target" co-resident species, burrowing owls and black-footed ferrets. Below, owlets not ready for long flights huddle in their burrow; a ferret brings her young above ground on a July day for lessons in the skills of hunting.

ROD ALLIN, BRUCE COLEMAN INC.

J. PERLEY FITZGERALD (ABOVE) AND PHIL A. DOTSON

BOB SHEETS

Playful, social—and officially endangered—juvenile Utah prairie dogs in Garfield County may lose their town. They live on land newly subdivided for summer homes.

Community in cross section: Barking with forefeet raised, a black-tailed prairie dog signals his territorial rights while neighbors engage in feeding, grooming, and play. Bison graze on prairie grasses that, left to grow, would hinder the dogs' ability to spot enemies—such as the swooping red-tailed hawk. More dangerous

PAINTING BY JAY H. MATTERNES

as a predator, able to enter the relative safety of the burrow, the ferret chases a prospective victim. But in the dark labyrinth of tunnels, the prey may escape. Burrowing owls, whose eggs and hatchlings provide food for rattlesnakes and prairie dogs, will devour defenseless pups on occasion—as will the hungry rattler.

Rifle in hand, a Wyoming biologist brings in a coyote he shot from the air. For a study of pronghorns, researchers investigated coyote predation on fawns and found it important; other studies have found the coyotes in question subsisting mainly on jackrabbits and on carrion. Nevertheless, coyotes have borne the brunt of stockmen's enmity since sheep and cattle came west. Weighing 50 pounds at most and often hunting alone, the coyote preys upon the weak, or scavenges the dead. At right, three clean the bones of an elk carcass in the National Elk Refuge near Jackson, Wyoming.

RON KLATASKE (BELOW AND LOWER RIGHT)

Scalped from ear to ear—the mark of bounty hunting—a coyote hangs on a barbed-wire fence in Kansas. This state ended coyote payments in 1970; but other jurisdictions have not, and ranchers' pressure for bounties increased after a federal ban on poisons in 1972. Coyotes, however, still flourish. Once found primarily in plains country, they have extended their range in the west and into the northeast.

STAN WAYMAN

might deduce from it, saying that his predecessor Donald K. Fortenbery, another field biologist from Patuxent, had done the sleuthing. The prairie dog seems always to dig by day, the ferret by night. The prairie dog's labors ordinarily produce a mound, which helps keep surface water out of the tunnels. But the ferret, as Don saw by spotlight, "backs out of the burrow holding the soil against its chest with the front feet, then passes it to the back feet, and kicks it out behind." This usually makes a trench leading away from a hole.

We were looking at what may have been such a trench. But it could have been the work of a badger. We couldn't infer from this that ferrets were here. Now, a fresh plug with a hole through it would be a good sign that a sealed-in ferret had escaped—and a good reason not to spread poison here.

coyote

Annihilation of a town can mean not only the destruction of overlooked ferrets but also the displacement and perhaps the death of another resident, the burrowing owl. The ferret, with endangered-species status, gets some federal attention. The owl, however, has had its day of listing. Included in 1966, it was dropped in 1968 and has remained officially out of trouble ever since. But Kenneth O. Butts of Oklahoma State University, who has studied the bird in that state, calls its condition still "quite precarious."

Like so many compact habitats, the prairie dog burrow sustains a myriad of life. Black widow spiders may take up a post at the entrance, seeking insect prey. Ground squirrels and mice may be uninvited guests. Vacant burrows may attract weasels or rabbits, snakes, salamanders, and toads. Plowing under or poisoning a town means the destruction of a tiny world.

The relationship of prairie dog and ferret, says a report on predator control, "seems to be a rare and evolutionarily finely tuned mutual adjustment of a vertebrate predator with an apparently highly vulnerable prey." Knowledge of such prey-predator balances is, in the most literal sense, vital for the survival of animals in conflict with man. Hard-won observations like Con's may save the ferret from unintentional execution at federal hands, while the Bureau of Sport Fisheries and Wildlife tries to satisfy ranchers who want prairie dog towns poisoned and conservationists who want to save the ferret.

Not all towns are threatened by "damage suppression measures." Those Con showed me within the Badlands National Monument are protected by the National Park Service.

Yet life struggles in that 170-square-mile dominion of ashen land and wind-carved rock, and life often succumbs. The wolf, the grizzly, the cougar, all are gone. Human invaders began tipping the scales of the delicately poised ecosystem in the 19th century. Con took me to a grasslands oasis there, where we sat on a hill and looked down on a prairie dog town. The dogs

By moose tracks frozen into three feet of snow, a coyote sits howling in northern Montana woods. As the morning wears on, he will begin his daily hunting, flushing out rabbits, rodents, or other small animals.

quieted down and their *yips* of alarm echoed away. Then, on the opposite hill, Con saw something move. He lifted his glasses, looked for a moment, passed them to me: there, crouched beneath a bush—a coyote. We waited awhile for it to stalk the town. But whatever the patient predator was going to do, it would not do it in the sight of man.

The coyote, like the ferret, lives a perilous existence. It is not an endangered species. Neither is it an innocent bystander. It confronts man, and man kills it almost without restraint. In the haven of Badlands I had seen one that would live.

RALPH H. WILLIAMS

The coyote finds few such sanctuaries. In cattle country, though, it has one advocate. Driving back to Con's office in Rapid City, we passed a sign that reminded me of shoot-'em-up Westerns in which sheepmen fight cattlemen. "We still have a sort of range war out here," Con said, pointing to the sign, unfancy but emphatic: COYOTES AND RATTLESNAKES PROTECTED. COYOTES GET THE SHEEP. SNAKES GET THE HERDERS.

No other animal in North America inspires more impassioned controversy than the coyote. The debate seems destined to last as long as the coyote. And both its champions and its enemies, many of them sheepmen, agree this may well be forever. Already it has survived decades of attack on "varmints."

Settlers, farmers, ranchers, and trappers did most of the killing until 1915, when the first federal campaign of "predator control" began. It has never stopped. Congress originally assigned this mission to the Department of Agriculture, which created an agency called Predator and Rodent Control (PARC). In 1940, PARC was transferred to the Department of the Interior. Changes of this kind often command little more public attention than underground events in a prairie dog labyrinth—but they may be equally significant for the survival of a species. PARC carried on with its mission. In 1963 alone, its records showed the killing of 842 bears, 20,780 bobcats, 2,779 wolves, 294 mountain lions, 89,653 coyotes. They did not show how many "non-target" animals—from squirrels to eagles—were killed, for by now the controllers relied heavily on poison, principally sodium fluoroacetate, discovered in 1944 and best known as compound 1080.

To predator killers, 1080 seemed ideal. It was cheap (25 cents' worth could kill a coyote) and it dissolved in water. It could be injected into a bait, usually a chunk of meat, and remain lethal for months. Grain soaked in 1080 was scattered to wipe out rodents. In 1963, PARC set 39,910 traps, scattered 151,942 pounds of poisoned grain, put out 708,130 poisoned baits, and placed 64,921 "coyote getters." These were short lengths of pipe stuck in the ground—a coyote that pulled at the getter's meat-scented wick triggered a spring that propelled cyanide into its mouth or set off an explosive charge.

The getter seemed appallingly cruel to some; and public concern grew over the secondary victims of 1080, for an animal that ate poisoned bait would itself become poisoned bait for scavengers, such as vultures and eagles.

Responding to public concern, Secretary of the Interior Stewart L. Udall empaneled an advisory board of wildlife experts, none of them connected with the Federal Government, for the first authoritative study of PARC by outsiders. In 1964 the board, under chairman A. Starker Leopold of the University of California, published a report highly critical of PARC. In 1965 Secretary Udall abolished PARC and created the Division of Wildlife Services to "alleviate damage caused by wildlife."

People who had hoped for change did not get it. In 1966 the new division supervised the killing of—by its own count—91,389 animals (with 768,688 poisoned baits).

Then, in 1966, Congress gave Interior a new mission: protection of animals threatened with extinction. Now the Division of Wildlife Services toils to make some animals scarce; the Office of Endangered Species toils to keep some scarce animals from becoming scarcer. At times both units deal with the same animals, such as wolves and mountain lions.

Again, in response to public pressure for something more than a change of name in predator control, Interior recruited an advisory committee in 1970. Dr. Leopold was a member; the chairman was Dr. Stanley A. Cain of the University of Michigan. The Cain committee added up the official toll taken by federal agents from 1937 to 1969. Bears: 23,803. Bobcats: 477,194. Wolves: 51,857. Mountain lions: 7,264. Coyotes: 2,823,146. It noted many secondary victims of 1080, from pet dogs in Colorado to a California condor.

"No method of control," said the Cain report bluntly, "is tolerable that threatens any species with extinction." It noted the disparity between the $8,000,000 spent each year for killing animals and the $50,000 estimated for aiding the imperiled eastern timber wolf, kit fox, red wolf, and black-footed ferret. It noted the "unenviable task" of protecting the last two and protecting livestock interests at the same time. It declared that the concentration on "substances and devices for killing animals" continues "largely on a basis of unvalidated assumptions."

Among these assumptions the Cain report placed the idea that widespread poison would control predation; it recommended a ban on such poisoning. It was submitted in October 1971. On February 9, 1972, President Richard M. Nixon by executive order prohibited use of toxic chemicals for predator control on federal lands, permitting it for rodent control. He also banned federal participation in any use of predacides. The federal war on animals no longer sanctioned unlimited chemical warfare.

But the animal wars go on. Reporting on them, I feel like a correspondent bombarded by conflicting communiques. Undisputed facts about predator damage are even more elusive than the varmints—including the black-footed ferret.

Sheepmen say the poison ban may wipe out sheep raising in the West. Since 1942, when 49,300,000 sheep dotted the rangelands there, the flocks have been dwindling steadily. In 1972, the number was down to 15,700,000. More than four out of every ten of those sheep graze at least part time on public lands for a fee that averages 20 cents per sheep per month. (The Cain report cited this as a form of public subsidy, along with

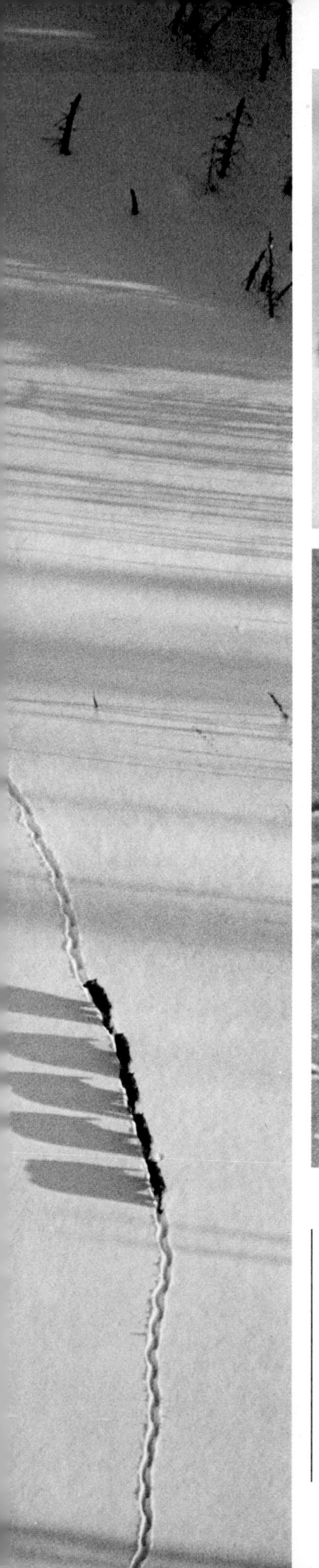

DAVID HISER (BELOW AND LEFT)

GORDON C. HABER

Traveling single file behind their pack leader, a pattern especially common in deep snow, wolves in Minnesota cover many miles searching for prey. Circling an old bull moose in Alaska's Mount McKinley National Park, a pack sizes up its chance for a kill. To identify the most dangerous adversaries, wolves rush at their prey, barking, testing reactions. Attacking a moose, they set their teeth in hindquarters and nose, hanging on and preventing blows of front hoofs. Having misjudged this bull, the pack finally gave up the battle. Formidable hunters and sociable companions in the wild, wolves reared in captivity—like the four-year-old above—often grow up docile and greet human friends as packmates.

JOHN H. GERARD, NATIONAL AUDUBON SOCIETY (ABOVE); JEN AND DES BARTLETT, BRUCE COLEMAN INC.

Sighting on the brilliant plumage of his state bird, a South Dakota sportsman and his dog anticipate the kill of a ring-necked pheasant, a favorite game bird since its introduction from Asia. Some hunters, blaming the fox for drops in the pheasant population, campaign to eliminate it; a five-year comparative study by the state finds minimal fox predation on the pheasant. Postmortems indicated that foxes more often consume mice or rabbits than birds.

predator control, protective tariffs, and import quotas; and it attributed the decline of the industry to competition from livestock reared in feed lots and from synthetic fibers.)

Sheepmen cite predator damage as a major factor in the troubles of their industry. In 1971 the Division of Wildlife Services concluded that predators had accounted for "nearly one-fourth of all losses to sheep." The Cain committee found "no basis for accepting these figures," but "no accurate source" of others for making an objective evaluation.

I asked a sheepman in California what his experience had been. He told me his losses ran between 4 and 6 percent in the past, but were climbing annually "and could go to 25." Sheep industry sources directed me to Maynard W. Cummings, a wildlife specialist with the University of California's agricultural extension service. He cited a report that in 1970 coyotes caused 56 percent of all livestock losses to predators within that state. Reported losses of sheep to predators have ranged from 3.6 to 14.9 percent in different states and different years.

Complicating any report is the fact that a sheep may die—from any of 25 infectious or 20 nutritional diseases, from infestation by parasites, toxic plants, birth defects, or bad weather—and appear to be a kill if a predator scavenges the body.

HENRY E. BRADSHAW

"If we are entitled to graze on public lands," argued Vern Vivion, president of the National Wool Growers Association, "we are also entitled to federal protection from predators." He spoke at the 108th annual meeting of the association, convened in Washington in January 1973.

This meeting touched off a battle over the poison ban, between sheepmen and environmentalists. The confrontation was especially significant because for the first time in its long career as a predator the coyote had found ardent defenders. It had become a symbol of mass poisoning. In a rare display of cooperation, 40 humane, conservation, and environmental organizations—some pro-hunting, some anti-hunting—united. They urged President Nixon to continue the ban.

Sheepmen responded with emotional as well as economic arguments. Secretary of Agriculture Earl L. Butz told the convention that "the public may eventually have to make a choice between lamb chops or hearing the howl of coyotes."

As the sheep-coyote issue became bio-political, Mr. Vivion, a Wyoming sheep raiser, said, "With our backs to the wall, we are regrettably forced to develop bounty programs as an aid to reducing coyote numbers." Thus the industry turned to an ancient remedy that many say is no remedy at all. The Cain report, and a study by Maynard Cummings that criticizes it, agree on one thing: the futility of bounties. And both, like Mr. Vivion's speech, call for research into ways of controlling coyotes by more scientific means than poison spread broadcast.

Once the link of prey to predator was thought to be as starkly simple as tooth or claw tearing food. Scientists now know that the relationship is deep and devious. But funds for research are inadequate. Dependence upon bounties appears to be gaining, although bounty hunters often concentrate where predators are abundant, not necessarily where they cause damage to stock.

H. Charles Laun, of Stephens College in Columbia, Missouri, maintains a "Bounty Information Service." Since the poison ban in the United States, he said, wool-growers were planning to pay $15 to $25 per coyote. He noted that when Missouri had a statewide bounty, most of the coyote payments were made in counties on the Kansas border. "Are there that many more coyotes in the extreme western counties of Missouri?" he asked. "No, it's just a matter of simple arithmetic." The bounty paid in Kansas was $2; the Missouri bounty averaged about $10.54. He considers the entire bounty system not only an invitation to fraud but "biologically unsound as well."

While I was in South Dakota to learn about the black-footed ferret, I discovered another front in the animal wars. This time it concerned foxes, with battle lines drawn between state biologists and private citizens who had convinced themselves that the red fox was wiping out the state bird—and prime game bird, the ring-necked pheasant. One weapon was a free-enterprise bounty.

The pheasant, introduced from Asia, made its debut in Oregon. In 1914 it was brought into South Dakota to replace disappearing natives, the sharptail grouse and the prairie chicken; and until recently South Dakota led all other states in the harvest of pheasants. (This, even in a poor year, was worth $27,000,000 paid for items like licenses, gear, lodging, and food.)

When the South Dakota pheasant population plummeted from 10,000,000 birds in 1963 to about 4,500,000 in 1964, a private organization began a campaign to eradicate the fox. This non-profit corporation, Pheasants Unlimited (which has no connection with Ducks Unlimited), takes credit for the killing of more than 150,000 foxes. It uses three methods.

For its "tagged-fox bounty," several foxes are captured and released, each bearing a tag worth $250 to the hunter who kills it. (The state itself removed the fox from its bounty list in 1972, along with the bobcat and the lynx; the coyote is still bountied.)

In a "Pheasant Restoration Contest," members of 4-H clubs and chapters of Future Farmers of America compete for cash prizes. They earn points for each new pheasant reared; they also earn points for killing predators. In one year, 19 F.F.A. chapters reported destroying 1,583 foxes, 1,490 raccoons, 1,278 skunks, 555 badgers, and 55 coyotes.

"A community fox hunt," says one of Pheasants Unlimited's pamphlets, "combines all the fun and good fellowship of a husking bee or a barn-raising with all the excitement of a train wreck or a schoolhouse fire." Hunters get detailed instructions: "To get fox out of the den, use Larvacide (which is a grain fumigant obtainable at most grain elevators). Place a small amount on a rag and chuck it down the hole.... The first pup

PATRICIA CAULFIELD, BRUCE COLEMAN INC.

Florida panther

Captive set free in the Everglades, a young Florida panther lingers near human companions. Parks and refuges may contain 150—or 300—members of this seriously depleted, officially endangered race.

generally comes out in 15 minutes. . . . Use a .22 to dispatch him. . . . Remove each pup shot from the hole. . . . Following the gassing, set traps for the old fox, who will return to the den site later in the evening." One man got 25 cubs in one day.

Pheasants Unlimited, which calls the fox "the Red Menace," tells hunters that "every fox you get saves a dozen or more pheasants, and every hen pheasant saved to raise a brood means that many more pheasants for your hunting next year." The South Dakota Department of Game, Fish and Parks tells a much more complicated story.

Carl G. Trautman, a research supervisor for the department, reported its five-year comparative study of fox predation. Where only foxes were controlled, he found "no major effects on pheasants, cottontail rabbits, and small rodents, even though an average of about 83 per cent of each year's fox population was eliminated." Jackrabbit populations averaged about 136 percent higher than normal. (Postmortems of 1,176 foxes revealed that they consumed more rabbits and mice than birds.)

Where raccoons, badgers, skunks, and foxes were controlled, the average population differences ran "132 per cent more pheasants per year, 63 per cent more jackrabbits, 50 per cent more cottontails, and 18 per cent more small rodents."

The Trautman report reminded taxpayers that it would cost $30 per square mile to control fox populations effectively, $41 to control raccoons, badgers, and skunks as well. And the poisons required would be lethal to humans and livestock.

Robert A. Hodgins, then director of the department, conducted hearings to assess public demand for a massive predator-control program to increase the stock of pheasants. He concluded that "most South Dakotans believed that this type of action . . . would be detrimental to the ecological balance of the state, and, therefore, was undesirable." At one hearing he had said that his department was "charged with managing all wildlife in the state—and the predators are among this charge."

South Dakota does not stand alone in this new awareness of the predator's significance. From Maine, where coyotes have staved off bounty hunters, to California, where rare foxes have gained official protection, many wildlife managers are abandoning the old kill-'em-all tactics. No coyote or wolf or mountain lion can yet feel secure, but there are signs that the unconditional war against varmints is ending.

In more than 20 states, predator control has become a new sport. Thousands of animals die—reportedly, 84,900 coyotes and 45,900 bobcats in California in one year. But shooting does not threaten non-target species, as poison does.

Montana, "striving for a completely new look at the animal control situation," in 1972 issued a manifesto on predators: "The question is no longer how to kill, sterilize, or otherwise simplify the ecosystem for our convenience. . . ." It considered that by the "preponderance of evidence . . . predators have an acceptable and proper place within all animal populations; they are not only tolerable but, very likely, essential members of any animal community."

Treed by redbone tracking hounds, a mountain lion snarls at bay. She wears tags, tattoos, and a radio transmitter in her collar—apparatus used by Maurice G. Hornocker and Wilbur Wiles to study lion habits in the Idaho Primitive Area. Reaching to pull a lioness out for measurements, Mr. Wiles relies on the adequacy of a dose of tranquilizer. Risky as this looks, in nine years of work Dr. Hornocker found "little danger from the animals. The danger's from icy slopes and dead trees," hazards to tracking. Kneeling in the snow, he tunes the transmitter in a collar to a directional receiver that lets him trace the lion's movements. Range shrinking, numbers uncertain—6,500? or more?—the western lions rank officially as safe. Conservationists think they need greater protection as well as further detailed study.

WILBUR WILES

MAURICE G. HORNOCKER (ABOVE AND RIGHT)

In Michigan, the fisher, a predator related to the marten, had been vanishing while its principal prey, the porcupine, was increasing—and destroying timber by gnawing the bark. Fishers were re-introduced in the Ottawa National Forest. Porcupines began disappearing. Foresters shot 360 porcupines in 1960; nine years later, they shot fewer than 50.

Robert B. Brander, a U. S. Forest Service biologist, and David J. Books, then a candidate for a Ph.D. in forestry, studied the porcupine-fisher seesaw and perceived in one forest the entire panorama of prey and predator. "Ecologically," they wrote, "the fisher is valuable because it preys on medium-sized herbivores such as squirrels, snowshoe hares, and porcupines, thus helping to check their potentially rampant fertility. If the fisher's niche were vacated, other carnivores would attempt to fill it. Birds of prey, the cats, the coyote, and the fox, and other members of the weasel family could compensate for the loss of the fisher in some respects. Only the fisher, however, preys effectively on the porcupine, thereby cutting down on substantial timber losses where the rodent has multiplied unchecked. The ecological argument, then, also becomes an economic one."

Both arguments appeal to Tom Lasater, who runs a profitable cattle ranch about 85 miles southeast of Denver and looks on his spread as an ecosystem. He is no fanatic coyote lover. "When they started chewing on our calves," he told me, "we had to shoot a few of them. We shot ten coyotes and that was it. We're in the calf business." But, he said, "we won't throw 1080 bait out and kill everything on the horizon."

He equates his financial success with an ecological approach. "I think it's actually good business to cooperate with Nature," he says. "Since we've given up on all insecticides, everything unnatural, it's boiled down to the survival of the fittest."

Once we killed because of the short view—the view of a farm emerging from a forest, a ranch appearing on a prairie. Now the view is longer—of a continent full of people and their cities. If there are to be wild animals in that vista, they must have their niches, maybe for our good as well as theirs.

Some animals can make it on their own. "Don't worry about the coyote," a biologist told me. "Wherever we live, he can live. He's even knocking over garbage cans for food right now in Houston." The coyote at the garbage can may be a realistic part of the vista of today and tomorrow. But other animals, in narrow niches threatened by man, cannot adapt—how could the black-footed ferret find a substitute for the prairie dog? And what of the ones that cannot flee?

I was told of one species—tiny fish, trapped by time in a desert pool—that I would have to hurry to see. Unless man intervened on its behalf, a species thousands of years old had only a few months to live.

Immune to infection if wounded by bristling quills, a fisher circles a porcupine, biting its nose, alert for a chance to rip into the unprotected belly. Porcupines often damage valuable timber by gnawing bark, so foresters consider the fisher an ally well worth preserving.

1.10
1.20
1.30
1.40
1.50
1.60
1.70
1.80
1.90
2.00
2.10
2.20
2.30
2.50
2.60
2.70

6. Besieged Strongholds of Life: Disruption and Defense

From minnow to mastodon—
a chronicle of habitat disruption
across the centuries

I lay face down on a rock ledge, peering like a Gulliver at a world hardly bigger than a bathtub. In the pool of glassy water below me, inch-long fish darted behind pebbles, nibbled gossamer threads of algae, and herded their nearly invisible young through the vast shadow cast by my head. I stroked the warm water and watched the bold and curious ones flit between my fingers. I could see 20 or so. A few feet away, in the cavern beyond the ledge, were about 200—the rest of the species called *Cyprinodon diabolis,* the Devils Hole pupfish.

The pupfish live in a cleft of the desert and of time. Their tiny world is bound by sheer walls that rise 50 feet to the surface of Ash Meadows, a desert region not far from Death Valley. Sunlight streams down the narrow shaft only on summer days, only for hours. But in its brief passage it stimulates the growth of the life-giving algae on the submerged part of the rock ledge. In the shallows along the ledge, blue-hued males court the smaller, duller-colored females, which often lay their eggs in the fine gravel found only there.

The shallow water along the ledge is something the fish must have for survival—"their kitchen and their bedroom," Dean

Inches from oblivion—a gauge records the impending fate of a Nevada species, the Devils Hole pupfish. Ranger Dean Carrett apprehensively checks for a critical drop in the water level of the fish's only habitat.

WILLIAM R. CURTSINGER

WILLIAM R. CURTSINGER.
(FISH APPROXIMATELY TWICE LIFE SIZE)

Devils Hole pupfish (male)

Garrett calls it. A National Park Service ranger and a guardian of the pupfish, Dean took photographer Bill Curtsinger and me to the home of *C. diabolis* during what seemed to be their final days. Even as we stood on the ledge and looked down on these masters of survival, their habitat—quite possibly the smallest of any known vertebrate species on earth—was draining away.

Devils Hole apparently was formed thousands of years ago when the roof of a limestone cave collapsed, exposing a narrow slit with three towering walls. The fourth slopes enough to provide a steep path down. Water, always hovering at a temperature about 91° F., never growing stagnant, wells up from an extensive subterranean system.

The site has been under Park Service protection since 1952, when President Harry S Truman withdrew a 40-acre tract from the public domain and designated it a detached part of Death Valley National Monument. Most of the monument land lies in California; Devils Hole is just over the border in Nevada and technically as well as literally the fish reside in that state.

In 1969, a large tract around Devils Hole was acquired by a private corporation from the Department of the Interior's Bureau of Land Management. The Bureau promised the National Park Service a buffer zone around Devils Hole. The corporation planned an irrigation system to grow feed crops in the desert; it had drilled wells near the buffer zone and in 1968 it had begun heavy pumping on other holdings. The pool level dropped. Federal officials protested, asking the State of Nevada to delay applications for more drilling. Nevada refused.

By August 1971, the pupfish were like condemned prisoners awaiting execution. There was only one move left—go to court and seek a reprieve. The Justice Department asked a federal court in Las Vegas to order a halt to the pumping. The corporate well-diggers agreed to limit their use of certain wells until 1973. The pupfish's guardians had to go back to court in July 1972. "The United States claims rights," they said, "in and to the use of as much of the waters . . . as is or may become necessary for the preservation of . . . the Devils Hole pupfish."

When I visited Devils Hole, the wait for the judge's decision had lengthened into months—and the water level was low. Without a permanent injunction, pumping might resume.

Dean Garrett pointed to what looked like a partly submerged yardstick on a metal frame that supported a wooden box. The measuring stick—a gauge calibrated in hundredths of feet—gave a visible record of what was being recorded continuously on a hydrograph inside the box. Dean interpreted the squiggles on the graph, which registered even the infinitesimal tides in the pool. *Tides?* "Sure. You can predict them," Dean said, "using regular tide charts."

The gauge showed the water level at 2.8 feet, meaning 2.8

Eighty feet down in the water-filled limestone cleft near Death Valley, divers take a census of the Devils Hole pupfish. Wells and pumps, tapping ground water to irrigate feed crops, threaten to expose a gravel-strewn ledge near the surface where the small fish feed and spawn.

WILLIAM R. CURTSINGE

feet below a high-water mark indicated by a copper disk. (The higher the number, the lower the water.) I asked Dean if 2.8 was a safe level. "It's not quite critical," he said. "But anything beyond 3.4 and we're in trouble." Beyond 3.4, most of the rock shelf is exposed, most of the algae wither away, and the gravelly breeding area virtually dries out. Even now, a drop of a few inches would imperil his charges.

Bill Curtsinger had swum into the cavern to photograph the fish there. When he reappeared, he enthusiastically invited me in. I borrowed his face-mask and snorkel and slipped into the shimmering blue water. Near the surface, I could sense the seduction of beauty below—an ever-deepening blue with a hint of eternity.

I swam slowly to the edge of the ledge and looked down it. Again, a Gulliver's view: Fish flashed past my face-mask, passing along a slot in the rock that by their scale was a deep valley of boulders and algae pastures. Daily, for reasons still unclear, pupfish swim down into a darkening abyss—as deep as 85 feet—only to return home, to the blue shallows.

When I left Devils Hole, Dean and other friends of the pupfish were still waiting for the court's decision. Several weeks later, he wrote me a letter. The judge had granted a temporary injunction until the government and the developers could work out an equitable, court-supervised agreement. But Dean was still worried. The water had dropped to 3.6. . . .

I wish I could report that the Devils Hole pupfish will survive. Obviously they live, if not from day to day, then from crisis to crisis. And they have lived, against all odds, in a saga as old as Death Valley's. Their elfin realm took shape in times that created the immense realm of the desert. They are, in fact, survivors of a great race of survivors—the desert fishes, creatures of springs and streams and little pools in the arid valleys of the continent's southwest.

While the last glaciers of the Ice Age crept southward, some 50,000 years ago, this was a region of abundant rains. Mountain rivers draining into Death Valley fed a network of streams and lakes, where small fishes ranged: minnows, mostly, and cyprinodonts, ancestors of today's pupfish. The lush vegetation of the land offered food for giant ground sloths, mastodons and mammoths, elk and deer, and animals we would recognize as camels and horses.

About 11,000 years ago, the glaciers began receding, the rains slackened, the streams narrowed and grew shallow, the lakes shrank, the land began to dry up. The big herbivores vanished; the great predators like the so-called saber-tooth "tiger" and the big cat *Panthera atrox* followed them into extinction. Desert conditions were fully established 4,000 years ago, when the bristlecone pines of the White Mountains, 150 miles northwest of Devils Hole, were seedlings. And by this time, in most of the

ROGER W. BARBOUR (1, 2) AND BRANLEY ALLAN BRANSON

(1) watercress darter

(2) Okaloosa darter

(3) Pecos gambusia

(4) roundtail chub

(5) woundfin

For fishes with limited distribution (1, 2, 3) or special requirements like clear pools (4) or swift turbid rivers (5), existence may end abruptly: a dam rising, a highway widened, a spring silted or dried up.

region, fishes could find a habitat only in isolated springs and their outflow streams. Most of the minnows perished. But pupfishes adapted and survived.

In the relatively rapid span of a few thousand years they developed into different kinds of fishes, each superbly adapted to life in a harsh new habitat. Some lived in water chilled by the snows of the Sierra. Others flourished in pools whose temperatures soared above 100° F., in sparkling fresh springs, in water extremely alkaline or saltier than the sea. Changing from generation to generation, they developed into recognizable species with a number of subspecies.

A triumph of resilience, certainly; but what's important about the fact that one species of pupfish has more fins than another? We usually shrug off such nitpicking as a matter for specialists only. Yet speciation, the emergence of new forms from a single ancestral population of plants or animals, is a key to understanding evolution and the survival of all living things.

Modern work on the evolution of the desert fishes parallels the work of Charles Darwin in framing the theory of evolution itself. Scientists today look to desert habitats for clues such as those Darwin found among the Galapagos Islands in 1835. Exploring the archipelago, he noticed that the finch there was no longer a single species but 13. Each had adapted to its habitat in subtle ways. He recognized a "perfect gradation in the size of the beaks"—suited to the different foods available. From such observations grew the theory set forth in his epic work on *The Origin of Species.*

Like the Galapagos Islands, the springs and pools that sustain the desert fishes impose an enforced isolation. Moreover, they can support only a small number of individuals. Thus they provide ideal subjects for study.

Scientists have numerous questions about these species, which are among our most endangered. As civilization invades the desert, however, the species rapidly disappear. Between 1930 and 1970, at least five species and eight subspecies became extinct in the continent's southwest. "By the time we're smart enough to ask the questions," one biologist told me, "the species that could give us the answers are gone."

What use can we make of such studies? Are there practical reasons for saving these unspectacular animals? I put these questions time and again to the biologists I met. One especially thought-provoking answer comes in the words of Alfred Russel Wallace, who recognized the principles of evolutionary change independently of his great contemporary, Darwin.

"The modern naturalist," Wallace wrote in 1863, ". . . looks upon every species of animal and plant now living as the individual letters which go to make up one of the volumes of our earth's history; and, as a few lost letters may make a sentence unintelligible, so the extinction of the numerous forms of life which the progress of cultivation invariably entails will necessarily render obscure this invaluable record of the past."

That quotation was handed to me by Edwin P. Pister, a biologist for the California Department of Fish and Game, and an authority on desert fishes. I met Phil Pister at Bishop, California, in the Owens Valley, where the pupfish had found a comparatively generous range of freshwater marshes—and a major new challenge with the coming of the white man.

In most of the cyprinodonts' desolate realm there was nothing the pioneers could want. But in this valley, snows of the Sierra Nevada fed a river that flowed all year long. When farmers dug irrigation ditches for the fertile soil, the Owens pupfish took the change with apparent ease. They spread into the ditches, where they helped keep down mosquitoes.

Then the Owens River was tapped to supply water to a city more than 200 miles to the south—Los Angeles. Developers built dams, drained marshes, introduced new fishes. In 1948 Dr. Robert Rush Miller, working with specimens taken years before, formally described the Owens pupfish as a new species, *Cyprinodon radiosus;* in that year he reported it "very scarce," its habitat drastically diminished.

In fact, Phil Pister told me, "for a number of years, we thought they were extinct. My current boss, Bill Richardson, found a tiny population in 1956. Miller came here in 1964 and said, 'Let's go out and see if we can find any.' Finally, we found a small colony, right out in back of us here." We were standing in a barren stretch of valley north of Bishop, looking at a trickle of water called Fish Slough.

Phil went on to say how the colony was nearly lost again in 1969: "It was one of those 100-degree-plus August afternoons. The pond we had the fish in was down to practically nothing—about the size of a ping-pong table. We could tell that if we didn't get those fish out of there we were going to lose that entire species, literally in a matter of a few hours. So we went out there and with our hands pulled those fish out of the mud and put them in buckets. . . . That's how close we came."

Phil took me to a little pond behind a low earthen dam. "The Owens Valley Native Fish Sanctuary," he said proudly. In the marsh-fringed, six-acre pool Phil had planted the Owens pupfish. "We're glad the City of Los Angeles is the landowner. This is the type of use they like to see. They gave us the land free, also the engineering and design work. State prison inmates built the dam and did a nice job." As we walked near a stand of willows, we flushed a few teal. "It turns out to be a nice area for waterfowl, too. A fish gate keeps out exotics—introduced species that could kill off the pupfish."

I asked how he got rid of the exotics in the first place. He winced. "We used chemicals; and in treating the water we seriously depleted a small snail population we didn't know about. We got a letter from a malacologist down in San Diego, madder than a hornet. The lesson is obvious."

A founder of the Desert Fishes Council, formed in 1969, Phil deals with a number of obvious factors. About 50 species and subspecies are at risk; some populations could be wiped

CHARLES J. FARMER

On her way to the Bridger Wilderness Area, Kathy Farmer paddles up Lower Green River Lake in Wyoming. Some 15 miles away, the Kendall Warm Springs, isolated from the river by a ten-foot waterfall, bubble out of the earth at a hospitable 85° F. In these inviting waters—and nowhere else—lives the Kendall Warm Springs dace, two inches long at full growth, now threatened by a proposed dam that would inundate the springs. Translocation to some suitable pool might offer hope; meanwhile the tiny fish need protection from anglers seining the waters to collect bait.

out by a single pump or the introduction of exotics ranging from goldfish to largemouth bass.

One remedy seems obvious: simply move the threatened fish to another spring or pool or tank. In 1972, thirty of the Devils Hole pupfish were put in plastic bags and taken by helicopter to a "refugium," a 20-by-10-foot concrete tank at the foot of Hoover Dam. The new environment was carefully considered, with warm water chemically similar to that at Devils Hole.

But these are emergency measures, the Council's biologists warn, not really an answer to habitat disruption; ". . . long-term living in an artificial or new environment would inevitably change the fish themselves, and the species that nature has produced would be lost."

Just how nature produced these species would be worth knowing. Apparently the pupfishes have evolved at an unusually fast rate, with an unusually high amount of genetic change. Mutations, more often than not, do not favor survival; and today an increasing exposure to radiation may increase the rate of mutation in human beings. We just might learn some lessons in the genetics of human survival from the pupfish.

In the tangible problems of land use and water allocation, Phil Pister sees a lesson. "For what we have in the Ash Meadows area, and throughout the Death Valley System," he says, "is far more than a few pupfish and a declining water table. We have in this system a tiny microcosm which reflects, basically, the same problems which the entire Earth faces. . . . Man must begin . . . to decide where the line must be drawn between environmental preservation and economic development."

We seem to understand about drawing that line when we are protecting spectacular, well-publicized animals. But when the line is hazy or not apparent at all, the chances are that the only persons who will see it are the scientists close to the ground, the ones who crawl around in search of their unsung, unknown charges. From those scientists I gained a new respect for the ugly ducklings of endangered species lists—the snakes and lizards, and frogs and toads, the salamanders and snails.

"There is no doubt," a federal official admitted to me, "that certain animals—glamor animals—have appeal. If the World Wildlife Fund sends out a plea for money for something like the red wolf, they'll get it. But if they tried to do it with the long-toed salamander or something, they couldn't. Callous as it may sound, I think we're going to have to concentrate on these glamor species because this is what the public wants. And if we don't, sooner or later Congress is going to say what the devil are we doing, and the whole ball game will be over." He paused for a moment.

"The answer may be to try to educate the public. But even the people who are concerned about the glamor species are in a minority."

Well, I have been educated by champions of the unglamorous, and I feel compelled to pass on what I learned. The so-called unglamorous species have much to tell us, once we're wise enough to ask the questions. For example, the translucent eggs of amphibians have been called "windows on embryology," through which we can see the marvels of growth. Snakes and frogs contain so many questions and answers, in fact, that they have become prime subjects for biology classes.

Ironically enough, this may pose a threat in itself. Leopard frogs and garter snakes are particularly popular as specimens. Carol A. Scott, a wildlife biologist in Manitoba, estimates that 24,800 garter snakes were collected in the province in 1971. That same year collectors were paid 50 to 55 cents a pound for leopard frogs. One man sold nine tons—about 216,000 frogs. One company harvested twenty tons.

PATRICIA CAULFIELD

Key deer

Now the interlake area, around Lakes Winnipeg, Winnipegosis, and Manitoba, has become the frog-collecting capital of the continent. (The title passed from Minnesota, where decades of collecting had depleted the supply.) In 1971 a Canadian law licensed commercial collectors; in 1973 certain limits were imposed on them. Manitoba's yearly quota for commercial catches of leopard frogs was set at 50 tons.

Already five species of reptiles and amphibians are known to be rare in the province; and Dr. K. W. Stewart, a University of Manitoba zoologist, is concerned about a secondary effect: Leopard frogs are food for many mammals, birds, and fishes; if the frogs disappear, such glamor animals as the heron and the mink and the northern pike may decline in turn.

A shortsighted, glamor-dazzled view can be as dangerous to us as to the benighted animals, according to Professor Robert C. Stebbins, curator of herpetology at a museum of the University of California. In his office at Berkeley, he talked with me not so much about his favorite animals as about what we are doing to our planet.

"Most people," he said, "don't realize what a devastating effect we're having on the great variety of living things. Big chunks of habitat are being destroyed. I wonder how long man can go before he finally reaches a point where things fall apart generally."

I asked if he were suggesting that the obscure animals were like the canaries formerly used to detect gases in coal mines; when a bird died, a miner knew that he would be next unless he escaped, fast. Dr. Stebbins nodded.

"If the present population trends continue, there may be a big die-off—of *people*—in the 21st century. I've seen what has been happening to the canaries in many parts of the world.

"You see damage and you ask yourself if a certain number of critical species will be knocked out to the point where the whole living system begins to tear at the seams.

"All through evolution Nature worked to produce highly complex ecosystems. Granted that species died out. But new ones

Deer No. 6, one of some 400 collie-size whitetails peculiar to the Florida Keys, carries telemetry gear for a study of habits and habitat—a habitat that shrinks as vacation or retirement homes go up.

pronghorn
mule deer
Bison antiquus
Equus occidentalis
Megalonyx jeffersoni
Nothrotherium shastense
Paramylodon harlani
Mammut americanus
Mammuthus imperator
grizzly
Ursus sp.
Arctodus simus
Panthera atrox
mountain lion
Canis dirus

Camelops hesternus

Tapirus sp.

man

In this panorama of scenes at the La Brea tarpits in southern California 10,000 years ago, white circles surround six species that survive today. All the rest vanished in a great and mysterious dying off of large terrestrial mammals in the Pleistocene Epoch. Evidently, however, the bison, black bear, camel, and tapir closely resembled living forms except in minor details. Herbivores like giant ground sloths wandered into these tar seeps and died in the mire, luring carrion-eaters and predators to their own death.

Today's wildlife seems impoverished, set against the fossils preserved here. Animals worthy of superlatives—largest, strangest, fiercest—have disappeared. Teratornis soared on 12 majestic feet of wing; mastodons roamed the continent; dire wolves and lion-size cats prowled the land. Worldwide, more than 200 genera became extinct during this puzzling time; to date no single theory really explains why. Some scientists blame man the hunter.

Mammuthus columbi

Smilodon californicus

Teratornis merriami

California condor

PAINTING BY JAY H. MATTERNES

arose and there was an overall equilibrium. As present life systems become more simplified, we're going to have an increasing loss of stability. I'm thinking of some crop pest, some parasite, some disease-carrying organism. It could just take off — *whoosh!* That's the kind of situation that worries me."

I wondered how he, a specialist in so-called lower life forms, had come to worry so much about people and ecosystems. The answer: His specialty gave him a clear view of habitat disruption and outright loss of species. "Reptiles and amphibians are usually restricted to small habitats. They can't fly away. When something happens to them you can see it all — if you're paying attention." He cited as an example what happened during the building of a four-lane freeway near Santa Cruz in 1969. The California Division of Highways was told by the Department of Fish and Game that no species of concern to them would be disturbed by the construction. But a population of rare salamanders, discovered by one of Dr. Stebbins' students, was living right in its path, at a pond named Valencia Lagoon.

This almost-ignored species was the Santa Cruz long-toed salamander, a black, gold-splotched animal about five inches long. It breeds in ponds, where its larvae develop. Valencia Lagoon was one of the two places on earth where it was then known to exist; the other, also threatened, was a few miles away. The road-builders had drained the pond before they knew the salamanders were there.

Now they have built a temporary pond, larvae have developed there, and the state is trying to restore the original habitat. Years earlier Dr. Stebbins had successfully recommended the salamander for state and federal endangered lists, and he hopes this species has been saved.

Dr. Stebbins was a kind of godfather at the christening of another salamander, one left in limbo by the "red book" errata sheet. This is a rather hapless creature — "an earthworm with legs," someone dubbed it — that has trouble capturing prey, suffers the ignominy of itself being the prey of mere beetles, and does not reproduce very well. But it has somehow managed to make a living for 12,000 years in a cool, damp forest about 8,500 feet up the north slopes of the Jemez Mountains in New Mexico. The trees themselves are survivors, remnant of a forest that greened the Southwest until after the last Ice Age.

The salamander apparently retreated with the forest, an animal that hid under rocks or rotting logs and remained unreported until 1926. The two specimens found then were to lie forgotten at the Smithsonian for 23 years longer. Then Dr. Stebbins and a colleague examined them and named a new species: the Jemez Mountains salamander, formally *Plethodon neomexicanus*. The plethodons, members of a genus found only in the United States and Canada, are evolutionary curiosities that have no lungs; they breathe through their moist, permeable skin.

GEORGE PORTER, PHOTO RESEARCHERS, INC.

Pine Barrens tree frog

Throat darkened in spring, a male Pine Barrens tree frog will call from a bough for a mate. Experts consider the species' limited habitat threatened, breeding potential in captivity "probably almost nil."

The habitat of the Jemez Mountains salamander is threatened by logging in the Santa Fe National Forest—and with the habitat, the salamander. Unglamorous as it is, it has gained the attention of the Forest Service and found champions in its own damp backyard. Their interest stemmed from a symposium on rare and endangered wildlife in the southwestern states.

Host for this meeting, held in Albuquerque in 1972, was the New Mexico Department of Game and Fish; its sponsors were the state chapter of the Wildlife Society and the state chapter of the American Fisheries Society. Participants included outstanding authorities on the fauna of the Southwest, an area drastically altered by man. I compared the long list of animals they discussed with the federal endangered list, and found a few animals on both: the red wolf, the black-footed ferret, the Sonoran pronghorn. Many of the less glamorous, however, were on the symposium's list alone.

The Gila monster, for one, the only known venomous lizard in the United States. Protected in Arizona for many years, the Gila monster became the quarry of collectors in New Mexico.

The ridge-nosed rattlesnake. Like the finches of the Galapagos and the pupfishes of the desert, these little rattlers of the mountains offer insights into evolution. One subspecies, for example, more closely resembles a race found 400 miles away than it resembles one found only 60 miles away—a puzzle for specialists in zoogeography, the study of animal distribution. Relative rarity has made this another species extremely valuable to collectors, and restricting collectors' permits does not in itself deter the unscrupulous. Arizona gave protection to the species in 1969; New Mexico has legislation pending for all its wildlife.

Scanning the symposium's list of birds, I noticed that some 165 species of songbirds breed in the Southwest (representing about 60 percent of all songbird species that breed in the conterminous 48 states). Of these, 52 "deserve local protection."

I paused over the section on prairie chickens. They belong to a genus—a group of species generally alike but distinctly different in detail—that is in its entirety "vulnerable to extinction." The name of this genus, *Tympanuchus,* bespeaks its best-known trait: the tympany of males dancing on the "booming grounds" as they court their females. From inflated air sacs on their necks booms a sound that once echoed over the prairies, audible a mile away. But the prairie habitat has shrunk as plowed fields have spread. Some subspecies are legally hunted; one, the heath hen, is gone; another, Attwater's, is on the federal endangered list—its numbers were down to 1,650 in 1972. The loss of a genus is a far greater tragedy than the loss of several species within it, for the genus represents not letters but words or phrases from the past, a biological fountainhead that wells from the diverging springs of life.

At the Albuquerque symposium, one of the participants was Roy E. Tomlinson, a field biologist assigned to the Southwest by the Patuxent research program. If you view the struggle to save our endangered species as a battle, then Roy Tomlinson in Arizona, like Con Hillman in South Dakota, is on the front line.

Unloved and often unnoticed, many reptiles and amphibians face the same relentless pressure on their living space as the more popular and better-known mammals or birds.

A few can sustain the role of living toy; at right, a girl tries the gait of a desert tortoise at the Arizona-Sonora Desert Museum. Even this creature has a prospective menace in off-the-road recreation vehicles churning up the terrain.

Too rare for casual handling, the San Francisco garter snake—one of North America's most vivid serpents, one of two on the endangered species list—has yielded many of its marshy haunts to highways and housing.

The Gila monster—with one Mexican relative, the only poisonous lizards in the world—lives so secretively that it languishes on a "status-undetermined" list.

The Jemez Mountains salamander, isolated in a small area by the retreat of ancient forests, has probably remained stable in numbers throughout historical time. Logging could destroy its margin of safety within months.

Similar predicaments make long-term survival doubtful for bog turtles, in scattered colonies from Connecticut to North Carolina, and California's black toads, found only in Inyo County. These two species have very limited ranges, exist in limited numbers; a relatively minor push can tip the balance against them.

DAVID J. DUNAWAY

black toad

DAVID R. BRIDGE, NATIONAL GEOGRAPHIC STAFF

desert tortoise

NATHAN W. COHEN

San Francisco garter snake

ROGER W. BARBOUR

bog turtle

JOSEPH W. SHAW

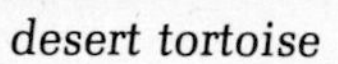

Jemez Mountains salamander

JEN AND DES BARTLETT

Gila monster

FRANK R. MARTIN

As members of a biological reconnaissance squad, they gather information and relay it to headquarters for use in recovery plans. There are exactly eight men in that squad. Two are working in Hawaii; one is in Puerto Rico. Only five biologists from Patuxent are assigned full-time fieldwork in the continental United States.

And work they do, night and day, year after year. They have been given a mission impossible, but they never call it that. When I met Roy Tomlinson, he was in the sixth year of a mission: the restoration of the masked bobwhite, a species that vanished at the turn of the century. Watching him at work, I learned more about the tortured environment of the Southwest than I could ever hope to absorb from the symposium papers or the O.E.S. recovery plans. And I could understand more fully the meaning of "habitat disruption"—that death knell for so many species.

In a desolate, grievously wounded land, searching with Roy for his phantom birds, I felt the same air of desperation that hung over Devils Hole. I would feel it in other places far from this arid valley in Arizona—the wild country of the condor, the Everglades' shriveling sea of grass. I would watch a biologist patiently trying to restore an animal torn from its ancestral home by man. One person dueling with one civilization and acting as if it were a fair duel. One person hoping for a miracle.

"Habitat disruption" is an abstraction that passes easily across your eyes when you read of a dying species. But stand now with Roy Tomlinson and hold one feather in your hand. That is all he has managed to find in a day of searching.

We are on a ranch in the Altar Valley, south of Tucson, near the Mexican border. We walk through the sparse grass, flushing mourning doves and jackrabbits, pausing at squat mesquites or bristling chollas. California poppies and yellow evening primroses daub the sandy soil. Roy stops at a cholla about four feet high and shows me one of the clues he has been looking for: a little woven basket, the abandoned nest of a cactus wren. Gingerly, he pulls it out through the needle-sharp cholla spines.

Roy examines the feather-lined bowl of the nest, smiles, and plucks his treasure. Holding the tiny feather between thumb and index finger, he evokes a masked bobwhite: "See the little after-shaft under the main feather shaft itself? Gallinaceous. Around here that means a quail. Chestnut-colored, so it came from the breast of a male—maybe from a dead one."

Later that day we tramp about a rocky patch of ground east of the valley, about 4,000 feet above sea level in the Sonoran Desert. We are on public property, administered by the Bureau of Land Management. The Bureau has staked out a "bobwhite study area" and fenced it off from cattle, which graze here for a fee, paid by a rancher who lives nearby. The bobwhite need

JOHN TOEPFER

C. ALLAN MORGAN

Special rites precede mating for the sage grouse (left, below) and the greater prairie chicken: the males' display of inflated neck sacs, and their competition for territory. But the nation lives on the food this land will provide; plow and harrow destroy the nesting grounds.

the scarce grass for cover from predators. Roy spots "trespass cattle" on the prohibited area: perhaps two dozen cows and calves. Yelling, running, herding them between us toward a gate, we try to drive them out. They scatter and placidly re-form in our wake. I can appreciate why cowboys use horses. We pay a call on the rancher, Roy has a friendly chat, and we depart. We didn't even see a feather this time.

The places I visited with Roy are two of three areas where he has released masked bobwhites in a long, complicated, and bold campaign to bring back to Arizona a bird that had been destroyed there—by cattle.

In 1870, there were 250,000 cattle in the Arizona Territory. In 1880, the railroad reached Tucson. In 1884, Herbert Brown of Tucson, staffman of the *Daily Arizona Citizen,* published a short notice on the discovery of a new bird, a bobwhite with a red breast and a black head and throat. In 1890, there were a million cattle in Arizona, and thousands of square miles of overgrazed and trampled grassland were rapidly becoming a wasteland. In 1897, what now seems to have been the last masked bobwhite seen in Arizona was presented to Herbert Brown.

Bobwhite still existed south of the border, where livestock did not become big business until the 1940's and '50's. But by the time Roy began his search for the bird in Mexico, in 1967, it seemed nearly extinct.

On a ranch in Sonora he finally found a population that was accessible, and not in danger. After studying the birds for months, he learned enough to convince himself and the Mexican Government that a small number could be trapped safely and humanely for a propagation colony.

By 1969 there were 200 masked bobwhites living and breeding in Patuxent pens. In March 1970, Patuxent shipped 160 banded birds to Tucson. They were taken to a promising habitat, watched in holding pens for 24 hours, and released. "They really seemed to love it," Roy recalled. "But they were dumb." Within two months they were lost from sight, most of them apparently the victims of hawks, owls, foxes, or coyotes.

Since then, the transplants have been given a longer period of adjustment, one to four months, and their release delayed until breeding time. Roy has concluded that the birds are breeding in the wild. But his campaign cannot yet be called a success. "We have to work out ways to teach them to be frightened of predators," he says. "We may even have to teach them what seeds to eat."

I asked why not just trap wild birds in Sonora and release them in Arizona. "Well, for one reason, we don't want too great an impact on Sonora's small population."

Since I saw Roy, he has completed his study of another endangered bird, the Yuma clapper rail, which nests along the lower Colorado River. Three rail subspecies are on the

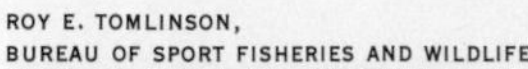
ROY E. TOMLINSON, BUREAU OF SPORT FISHERIES AND WILDLIFE

Yuma clapper rail

masked bobwhite

NATIONAL GEOGRAPHIC PHOTOGRAPHER OTIS IMBODEN

Three against the odds: Roy Tomlinson, replacing a masked bobwhite in its cage after weighing, fights a lonely battle to save it and the Yuma clapper rail from the effects of overgrazing and of marsh drainage.

endangered list. They are among the birds that need wetlands, wildlife's most endangered environment.

When a marsh is drained, when a bay is filled, the gain for man is usually far more obvious than the loss for wildlife. The animals' epitaph is short, unemotional: habitat disruption. A massive loss is far less visible than a small one. We can see a Devils Hole. But when many species make a last stand over an entire region, we cannot see them dying. Then we need the eyes of biologists. And often we need their voices, crying in what used to be the wilderness.

I heard the most urgent cries in Florida, whose wildlife includes 12 animals on the endangered list. Hawaii, with 29 endangered species or subspecies mostly found nowhere else, or California with 22 and Texas with 17, would seem to have more animals in trouble. But Florida's native species are being overrun by invading animals. "South Florida," says a report by two biologists there, "is becoming more like South America, Southeast Asia, and Africa in flora and fauna every day. For each introduction one or more native forms must give up some food, territory, freedom, and, possibly, continued existence." Another report calls the whole state "a biological cesspool of introduced life."

Introductions, or exotics, have been arriving in North America ever since the white man began introducing himself to various sections of the continent. Florida's exotic invasions began in 1884, when, according to one story, a lady bought a beautiful South American aquatic plant at the Cotton Centennial Exposition in New Orleans and brought it back to her home on the St. Johns River. The lavender-flowered plant was the notorious water hyacinth, which has blanketed the rivers, streams, and canals of the state. Attempts to kill off this and other water weeds have cost some $10,000,000 a year and, in many places, set off devastating ecological chain reactions.

Florida's thriving floral transplants—more than 1,000 species of exotic trees alone—form the backdrop for a drama that pits native fishes and birds against a host of invaders. Some native fishes are already in a rout.

Dr. Walter R. Courtenay, Jr., of Florida Atlantic University in Boca Raton, peered at me over an array of bottles containing exotic fishes that he and other workers had collected in a five-year study supported by the U. S. Fish and Wildlife Service and the Florida Game and Fresh Water Fish Commission. "We've just finished itemizing what's there," he told me. "Now we're looking at the impact. And it's pretty scary."

At least 20 species of exotic fishes and 5 hybrids are "established as reproducing populations" in fresh and brackish waters. The exotics are rapidly taking over the habitats of natives. Within the next two decades, says Dr. Courtenay, the most successful of the invaders, perchlike tropical fishes called cichlids, "will be the dominant fishes in south Florida." The losers will be the sunfishes, which are more American than apple pie. (The sunfishes—such *(Continued on page 147)*

From habitat to homesites, one hillside in Virginia undergoes change. In a glade of mixed pine and hardwood forest, sound or odor warns of man—a doe and a buck leap to cover, white hairs of tail and rump flashing a silent alarm. A fox and a black bear sniff for scent on the air; a squirrel perks its ears.

By late summer, tree cutting has revealed the Blue Ridge foothills; silt settles in an artificial lake; earth mover and bulldozer scrape away topsoil as they mold the landscape for a recreation community complete with its own golf course and tennis courts—and, perhaps, a few squirrels and opossums eventually.

Every year since 1950, man the consumer has taken 750,000 acres or more for economic purposes, with the speed and whine of a chain saw ripping through heartwood. The good life enjoyed by millions brings damage and death to the natural fabric of the continent. But even the concept of habitat disruption implies conflicting human interests with no easy reconciliation in sight.

PAINTINGS BY JAY H. MATTERNES

Block Island meadow vole
Ozark cave fish
Shasta salamander
blunt-nosed leopard lizard

The Lesser Legions

The wolf, the elk, or the bison—when only a few of such a conspicuous species remain, concerned people may at last rise to defend them. But often for lesser creatures mere existence passes generally unremarked.

Often they consist of a locally isolated subspecies, while the larger group does well over most of its range. The Block Island meadow vole falls in this category. A single population restricted to some 600 acres on one island off the New England coast, it thrives on vacant lots among beach grasses. Every house, road, or dairy farm built shrinks its tiny domain.

In California's San Joaquin Valley, conversion of grassland to agriculture leaves the blunt-nosed leopard lizard, by expert assessment, "on the verge of total extermination." Farther north, the Shasta salamander lives in a region of limestone subject to commercial use; the female lays a clutch of about nine eggs, each with two jelly-like spherical envelopes, but breeding potential in captivity remains unknown. Because of this species' rarity, biologists covet specimens, further depleting their numbers—an ironic fate that may also befall the blind Ozark cave fish, found only in a few caves and wells in Missouri and Arkansas.

Erosion of Canada's Sable Island, only breeding ground of the Ipswich sparrow, and development on the Atlantic coast, where it winters, reduce the numbers of this plain little bird.

No one, perhaps, but the specialist would miss any of these creatures. Yet each life form plays a unique part in the intricate web of the environment, and demonstrates a unique adaptation to the conditions of its world.

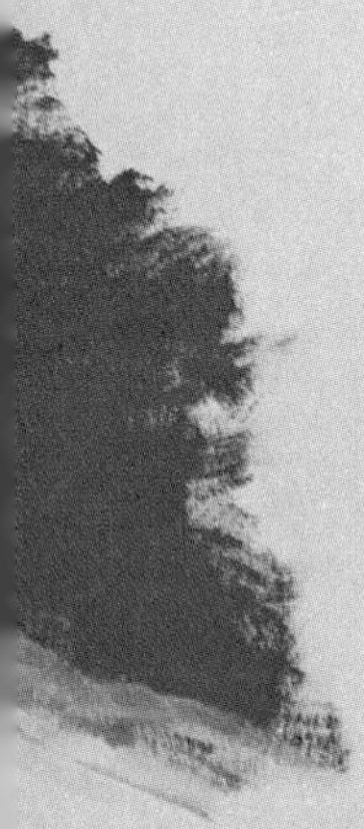

PAINTING BY JAY H. MATTERNES

Ipswich sparrow

On an oil-stained beach north of the Golden Gate, a young Californian scans the surf for birds in need of rescue—casualties of a tanker collision in 1971. Such accidental spills wreak ugly damage on prized shoreline, long-term injury on local ecosystems.

Across the continent, coastal wetlands recede; developers build more than 2,000 homes in New Jersey's Manahawkin Bay. Since 1945, many thousand acres of marsh and dune by the Atlantic have changed form as buyers compete for waterside property.

TOM MYERS

LAURENCE LOWRY, NATIONAL AUDUBON SOCIETY

angler's favorites as bass, bluegill, and crappie—are natives; the apple is an import from Europe.)

The fishes in his little bottles are displays in an ichthyological chamber of horrors; for each he has some grim data.

The walking catfish, for instance. Able to breathe air and slither along using spiny pectoral fins as "elbows," this Asian native was first spotted near Boca Raton. It now infests Lake Okeechobee and threatens Everglades National Park. "It's capable of displacing valuable game fishes like largemouth bass."

Or the black acara. A cichlid from South America, it is now in at least five counties. "Sport fishing in Florida could be wiped out by this one lousy exotic."

Or the piranha. This terror of South American waters—a school of them can strip the flesh off a cow in minutes—is not established in Florida, but.... Two, kept in an abandoned swimming pool in Miami, survived one of the state's coldest winters. And a Florida highway patrolman caught one on a hook and line in a pit filled with water from a canal.

Most of the exotics arrived by way of fish farms that raise stocks for home aquariums, and escaped in overflows or via drainage pipes or simply got dumped into canals. "The people in the aquarium industry," Dr. Courtenay says, "are mostly responsible people. I'm convinced they didn't know what they were doing. Now many of them are genuinely concerned. Canals connect everything from Miami to Orlando, and the interconnections run through the fish farm areas."

The Exotic Fishes Committee of the American Fisheries Society, consulted by the Department of the Interior, urged in 1973 that a number of species be barred from the United States. The aquarium industry endorsed the idea of a banned-fish list. But some entrepreneurs feared that a campaign against exotic fishes could destroy the industry—and a hobby pursued by about 20 million people.

After hearing some of the problems that exotics import with them, I was surprised to learn that state and federal officials were contemplating the release of another one, females only. This is the grass carp—also known as the white Amur, after the Amur River of its homeland, on the Soviet-Chinese border. The carp, which can grow to more than 70 pounds, is a plant-eater; a young fish can consume several times its weight in weed a day. But it has a very short intestine; one biologist wryly suggested putting diapers on the fish to prevent reinfestation by fragments of undigested weeds.

Exotic fishes may indirectly threaten the native fish-eating birds of the Everglades, already under new pressure from exotic birds that introduced themselves. An authority on birds of southern Florida, Dr. Oscar T. Owre of the University of Miami, told me bluntly: "The peninsula of Florida as we have always thought of it is simply not here today. Goodness knows what *is* here. We've turned into an ecologically disturbed area. We should have worried when the cattle egret arrived."

The cattle egret apparently came from the Old World, via South America, in the 1940's. In 1971, a state biologist reported that it is probably more numerous now than all the native species of herons and egrets combined. Dr. Owre adds that cattle egrets invade the roosts of other birds, such as herons: "They're aggressive and take over nesting space from other species."

Dr. Owre keeps study skins of his exotics laid out in trays. As he showed them to me, he began recounting bird facts that depressingly echoed Dr. Courtenay's fish stories.

The canary-winged parakeet, for example. "About 700 of them are roosting right in front of Miami city hall. They're already causing backyard fruit damage. They seem to be sticking to the urban area—hawks don't come into cities." Or the monk parakeet. "We're told it's one of the most serious agricultural pests you can get. Anyone who knows anything about parrots realizes that they pose a threat to Florida's agriculture."

The red-whiskered bulbul, probably released by accident from a tropical-bird farm in 1960, is increasing its numbers at an annual rate estimated at 30 to 40 percent. Dr. Owre worries about the chance that this cardinal-size import from Asia may threaten the state bird itself—the mockingbird.

"Our feeling is that the bulbul here has a diet almost identical to the mockingbird's. Now, male mockingbirds are intensely territorial in winter. The fact that he defends a territory guarantees him food. But the bulbuls in winter are gregarious. He can't drive them out. With them, he has a potential problem."

Dr. Owre puts some of the blame for the release of exotics on "people who are liberating birds deliberately, with the idea of 'saving' them from extinction in their native habitat. There are also hundreds of outdoor aviaries and zoos. All that some birds have to do to escape is fly up and away."

Virtually all of Florida's aliens have been uninvited—including such bizarre pests as a poisonous giant toad and a voracious fist-size snail that even ate paint off houses. But some, like the grass carp, had invitations. For several months in 1971 two Florida agencies fought an intramural battle over an import, the Game and Fresh Water Fish Commission against the Department of Natural Resources.

The game commission proposed placing some 40 rosy-billed pochards, South American ducks, in a state preserve near Gainesville, for an elaborate study to determine whether the pochard should be introduced as a game bird. The department, which had jurisdiction over the preserve, rejected the proposal.

Imported from Argentina, the birds were already in Florida, as were more than 60 other birds of various imported species. A similar selection was being held in Louisiana. Such studies were conducted by interested states under a federally supported program now extinct. As game species, these were officially endorsed exotics. They had sponsors.

In bio-political terms, it would seem that there are just two species: the sponsored and the unsponsored. Can endangered animals, as well as exotics, be classified this way? I asked myself that question as I set out to learn about two native Florida species on the endangered list—and other species, sponsored, unsponsored, and in trouble.

SALLY B. MYERS

EDWARD M. BRIGHAM, JR., NATIONAL AUDUBON SOCIETY

Disruption by diffusion: Dusting an orchard near Sacramento with pesticide, a farmer fights for a share of his county's multimillion-dollar pear crop; as many as six sprayings a year combat such pests as aphids or codling moths. Wildlife perishes also—birds of 23 species died after an application of dieldrin (two pounds per acre) for Japanese beetles in Michigan.

7. Sponsors of Survival

Who decides an animal's fate? Letter-writers and lobbyists, hunters, tourists, lawmakers, voters

Woodie Hartman came out of the post office as quickly as he had entered. "Nothing yet," he said, slipping behind the steering wheel. He cast an anxious glance at the ominously low gas gauge and started up for the drive back to the cottage that bears the impressive title of Manatee Research Project Headquarters. Woodie was broke and, in the absence of any check-bearing mail, so was his project.

Dr. Daniel S. Hartman is a rare specimen of animal scientist. He is not employed by any governmental or academic institution. And he does not hide his feelings for the animals he studies: "It's difficult for me to understand how wildlife biologists can study higher vertebrates with such detachment. No emotional involvement. All you ever hear is 'sustained yield' or 'harvestable surplus'—as if the only justification for an animal's existence was its practical worth to man! An animal should not have to justify its existence."

The animal Woodie loves is the Florida manatee, whose alternative name, sea cow, suggests its bulky shape as well as its disposition. The manatee may have inspired sailors' yarns

Bristled and blotchy, sluggish of temperament, virtually defenseless, the Florida manatee lacks the organized citizen sponsors who guard the interests of "glamor animals," more popular threatened species.

JAMES A. SUGAR

JAMES W. LATOURRETTE

DAVID R. BRIDGE, NATIONAL GEOGRAPHIC STAFF

At the Seaquarium in Miami, a manatee feeds on glossy leaves of water hyacinth, named for its delicate blossoms and notorious as an aquatic pest. Floating mats of it clog the waterways of Florida. In 1964 it suggested a useful role for the manatee; a single adult, scientists calculated, could clear 100 cubic yards of weed a day. An abortive experiment to prove this left the manatee with only one active ally—freelance biologist Daniel S. Hartman.

Equipped for note-taking with a pencil and a plastic chart, Dr. Hartman swims beside a subject in the Crystal River, southwest of Ocala. Such forays form part of his effort to determine the present distribution of the manatee. From North Carolina to southern Texas these mammals once flourished in numbers sufficient to stock—very briefly—a trade in "sea beef." Coastal waters of Florida and southern Georgia form their known range today.

JAMES A. SUGAR

about mermaids. Any sailor who mistook a manatee for a siren must have been, in Woodie's opinion, delirious.

Still, there is something lovable about the manatee. Seeing its pathetically homely, bewhiskered face appear out of the water one day, I could sense what Woodie meant when he said, "I have always felt that they are so ugly as to be beautiful."

That quotation, incidentally, is from his testimony before a Congressional subcommittee, when he proposed a national manatee refuge. His campaign for a refuge has been as futile as was his quest for a check that day at the post office.

The manatees, meanwhile, are going off the earth for the same reason that television shows go off the air: no sponsor. Unsponsored species can, of course, hope for next season. There cannot be a hope for a rerun, though, if a species is canceled for lack of interest.

Woodie's manatee project lives on whatever funds he can find. His earliest work was supported by a grant from the National Geographic Society. Money has also come from the World Wildlife Fund, the Friends of the Earth Foundation, and the Office of Endangered Species, which has the manatee on its list. But, like many other imperiled species, the manatee lacks an organized band of supporters to sound the cry—and plead for the money. *Save the whale!* can summon a crusade. *Save the manatee!* summons a question: What's a manatee?

I got my answer the day I set out in Woodie's punt across the broad, spring-fed headwaters of the Crystal River, 70 miles north of Tampa. "When I first came here in 1967," Woodie said, "you could look right down—as deep as 40 feet—and see the bottom below you. The manatees were easy to spot then. Now look at it. You can barely see a foot." I asked him what had happened. "A plant explosion. Hydrilla, water milfoil, hyacinth."

We saw a few dark shadows as we slowly cruised the headwaters. And once, out of the turbid water, poked a manatee's head, which quickly disappeared. Woodie nosed the boat through a dense green arch along the shore. The water in the little cove was crystal-clear. "This is all that's left," he said. "And it's going soon."

Woodie, who has snorkeled with more than 70 manatees all told, whimsically gave many of them such southern names as Pearly Mae and Flora Merry Lee. He can recognize them by their scars. Nearly all have been deeply gashed by powerboat propellers. "That's why they need the refuge—or at least protection," he says. "Just a speed limit would help."

Slaughter of the manatee for "sea beef steaks" swiftly brought the sluggish creatures near extinction around the turn of the century. Once flourishing from North Carolina to the Texas Gulf Coast, they clung to life only in isolated backwaters. Their range today is a mystery Woodie and his two assistants are trying to solve. Ever watching the gas gauge when they have a car, hitchhiking when they don't, they interview commercial fishermen, guides, crabbers, baithouse operators, asking them when they last saw a manatee. In some places, the animal

NOEL F. R. SNYDER

FREDERICK KENT TRUSLOW

Florida Everglade kite

grew even more scarce within the past decade. Small populations are scattered along the coasts of the Florida peninsula, notably in the St. Johns River drainage system on the east coast.

Once, not long ago, the manatee had a sponsor. In the Bio-Science Building at Florida Atlantic University is a glass case containing the mounted body of Claire, an 11-foot manatee. She was one of eight enlisted in 1964 as "biological agents" in one of the innumerable attempts to eradicate water hyacinth. Each manatee, theoretically, could clear up to 100 cubic yards of aquatic weeds a day. Before the experiment ended, Claire and six others died, apparently victims of unseasonable cold. Nothing much has been heard since about practical uses for manatees. "By conventional definition," Woodie says, "they offer no promise of pleasure or profit." And so they wait, unsponsored, but not entirely unloved.

The Florida birds on the endangered species list have a sponsor, the United States Government. One of Patuxent's field biologists, Paul W. Sykes, Jr., is assigned to them. Like his counterparts in South Dakota and Arizona, he earns his living by studying imperiled species.

Woodie Hartman wears his hair long; Paul Sykes has a crew-cut and his lifestyle is ranch-house-suburban, not cottage-commune. But the two men are not that different. They both work hard to save their animals, and both see themselves as defenders of nature's embattled world. "For a long time," Paul Sykes told me, "I've looked upon myself as a part of the environment—not apart from it. I have tried to get a personal feeling for the environment. I know that what man has done he has done from greed and from ignorance." One thing man has done is all but eradicate a marvelous bird, the Everglade kite.

Paul Sykes has been working for six years to save the kite. It seemed to me that first he would have to save Florida. He is trying to do something like that—at least enough of the state to give the kite a chance to become once more, with Paul Sykes, "part of the environment."

He took me first to the Loxahatchee National Wildlife Refuge, 55 miles north of Everglades National Park. The refuge, a 220-square-mile remnant of Florida's unique and vanishing sea of grass, is bounded by leveed canals—a system that determines the water level in the saw-grass marshes of the refuge. A tap on this reservoir means more water for millions of people and less water for a habitat that supports a vast and varied community with at least three animals on the endangered species list: the kite, the alligator, and the Florida panther.

Until large-scale draining began in the 1920's, kites graced nearly every marsh in the state. An unusually gentle kind of hawk, the kite sailed alone over the marshes, taking a single species of snail as its only prey. The slow-flying kite, though no game bird, was an easy target; countless numbers died to give

Talons grasping an apple snail, its only food, an Everglade kite takes wing. At Lake Okeechobee, an adult male guards the nest he built. Drainage of Florida's freshwater marshes threatens predator and prey.

gunners the sight of a plummeting bird. Nearly all the rest, deprived of the habitat and the food to which their species had adapted, died out.

From a tower near the refuge headquarters we looked down on a 600-acre habitat being created for kites. A huge, diesel-powered mower chops through the marsh and into the substrata, rooting out dense vegetation in an effort to open up sloughs. Plans call for maintaining a continual water level over the "kite management area." More than 4,000 apple snails—the only food the kite will eat—have been scattered in the marsh like seeds. "If we can keep the snail population high," Paul said, "maybe we'll get some kites to come here. . . ."

Meanwhile, the kites take their chances on another expanse of marsh in the Everglades. To see them, we put Paul's airboat into a slough across the highway from the borders of the park. Within ten minutes we spotted our first kite. Paul raised his binoculars: "Adult or subadult male."

With our second sighting, we managed to get close enough to see a kite moments after it had captured a snail. I watched through the glasses while Paul provided a running commentary. The kite was perched in a dead tree over the edge of a hummock. "A feeding station. They fly head down, looking for a snail, descend slowly to grasp it with a foot, to snatch and eat it at a favorite perch." Something dropped. "The operculum. You know, that kind of trap door on a snail. Step one." The kite reached its slender curved beak into the shell. "Step two. The curvature of the bill corresponds to the shell's—so the kite can sever the muscle that holds the snail's body to the shell." The shell dropped. "You can often find lots of shells at a place where they feed." The kite finished its morsel, sat for a moment, then flew off, head down, looking for another snail.

I looked down myself and noticed a snail on a blade of saw grass just below the surface. I reached for it—and it dropped like a stone to the bottom. Getting one is not as easy as a kite makes it look. The snails, which grow to about the size of ping-pong balls, apparently live in a classic prey-predator balance with the kites. The birds do not eat the snails' eggs, which cluster like tiny pearls on plant stems. The kites hunt only those snails they find in sloughs; the rest, in thick vegetation, are safe.

I asked Paul how many kites he thought existed. He made "a crude estimate" of 70 to 100. He has banded 34 nestlings, an unprecedented achievement. He has found 42 nests. For each he keeps an elaborate record of size, location, type of support (usually a woody-stem plant), height above water, depth of water below nest, materials of nest construction and lining, nesting success (from eggs laid to hatchlings fledged), and predation (ants and snakes). The records are a kind of insurance policy; all such minutiae will be needed if the day comes when every kite must be "managed" to keep it alive. Tossed a few scraps of marshland by men, it might survive. But I know one man who will curse the greed and ignorance behind that day.

WILLIS PETERSON

Kaibab squirrel

MARTY STOUFFER

Delmarva fox squirrel

Morro Bay kangaroo rat

BURDETTE E. WHITE

Across the continent, I found a bird in even worse trouble. It may be the rarest species in the United States, and many authorities single it out as the most endangered of all. More than any of our animals, it embodies the dilemma of imperiled wildlife: Man threatens it with extinction; only man can save it. Its very name reflects another dilemma: *California*—modern, conquered, booming. *Condor*—ancient, untrammeled, dwindling. Can state and bird—can man and wildlife—coexist?

I know of no other North American animal that has put its would-be saviors to a greater test. The 50 to 60 surviving condors need for their continued existence a range measured in miles; they have chosen the nation's fastest-growing, most land-hungry state. They need almost primordial quiet and isolation; they have chosen an era of superhighways and jet-streaked skies. They need a type of food—carrion—becoming rare near the nesting sites they stubbornly cling to. Changeless in a changing land, they are Pleistocene relics, giant birds roosting in the wrong place at the wrong time.

But a relic can become a sacred object. That seems to be what happened to the condor, and that seems to be what is saving it. Its many sponsors—agencies of the United States Government, the State of California, private organizations, a host of individuals—are united in the belief that the extinction of the condor would mean far more than the death of a species. People in condor country seem to see their bird, soaring along the edge of eternity, as a haunting symbol of life itself.

I was not surprised to hear one of the condor's champions say, "Working to save the condor is doing the work of the Lord." The man who told me that is Sanford R. Wilbur, the fourth of the Patuxent field biologists I met. His assignment is the California condor, and he carries out his mission as a scientist and as a deeply religious man. He once set down his own answer to his own question, why should we save the condor?

"I know God had a good reason for making each species," Sandy wrote, ". . . and I know He gave Man stewardship over all His earthly creations. . . . My scientific training tells me each species of animal and plant has a special place. . . .

"If the condor goes, and its slowing and stabilizing influence is removed, I have serious doubts anything else will be able to stop Development in all its many forms from consuming the remaining good parts of Southern California."

The heart of condor country is Los Padres National Forest, which extends along the mountainous coast and encompasses 1,950,000 acres of the state's wildest land. Within it are two special condor havens, the 1,200-acre Sisquoc sanctuary and the 53,000-acre Sespe sanctuary. I looked forward to visiting these, but Sandy shook his head. "I'm sorry. Nobody goes into a sanctuary." *Nobody?* "Almost. Just three of us can go there, and we don't go there very much."

Scanty in numbers, restricted in habitat, each of these isolated populations finds protection in state law—the Delmarva squirrel in Maryland, the Kaibab in Arizona, the Morro Bay kangaroo rat in California.

The three men represent the three basic defense forces arrayed to protect the condor: Sandy, the Bureau of Sport Fisheries and Wildlife, Department of the Interior; Dean Carrier, biologist assigned to the condor by the U. S. Forest Service, Department of Agriculture; John C. Borneman, full-time "condor naturalist" employed by the National Audubon Society.

Sanctuaries as such are not the key to the condors' survival. These largest of North American soaring land birds forage over a vast, wishbone-shaped area, 50 miles by 400. Each branch is a natural corridor along which the birds sail on nine-foot wings, seeking food. But the great herds they once scavenged—pronghorn, tule elk, mule deer—no longer exist. They can nest in relative safety, but they have increasing trouble finding enough to eat.

"Everything bad that has happened to the condor, has happened because of man," says Sandy. He does not want people to think of the condor as some kind of living fossil. "There is no evidence of genetic problems. If they are on their way out, it's an artificial evolution that we have caused."

Sandy began seeking clues to the birds' apparent decline in 1969. His investigation followed that of his Patuxent predecessor, Fred C. Sibley, who in 1969 reported "a decline in nesting since 1940 and an annual reproduction insufficient to maintain the present population." Sandy has built a case that indicts people of the past, present—and future.

The past: Egg-snatchers may have put a heavy pressure on the condor. Sandy found collection records of 198 birds and 58 eggs. "I counted all those condor skins in museums," he says, "and I wished some of them were still flying around." Shooters took an unknown toll. The condor was unjustly accused of killing livestock. Also, prospectors liked the hollow quills of the birds' feathers as containers for gold dust.

The present: Habitat disruption—but never on such a scale. Sandy compared condor-country maps of the 1950's with maps of 1967 and 1969. He estimated that in a key "condor feeding area" of 200,000 acres, about 20,000 acres have been lost. Between 1960 and 1970, the human population in parts of condor country increased three- and four-fold. Citrus groves, oilfields, and real-estate developments are spreading across the condor's traditional foraging areas.

The future: Fairly bleak. Sandy's colleague John Borneman puts it this way: "The 'just leave them alone' approach to condor preservation will no longer suffice. . . ." In 1972 alone, he points out, vacation homes took 100,000 acres of forage area.

Exploring modern condor country, Sandy and I drove along the western feeding corridor, then crossed to the eastern branch of the "wishbone." We left the high, rainswept ridges of the

JOHN JAMES AUDUBON, PLATE 66 IN "THE BIRDS OF AMERICA," NATIONAL AUDUBON SOCIETY (OPPOSITE); MARTY STOUFFER

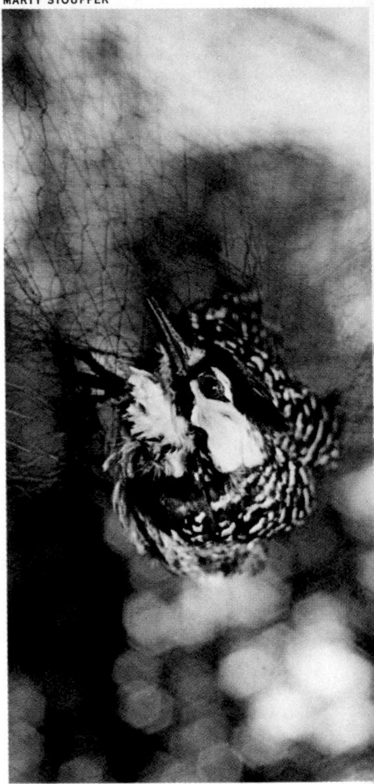

red-cockaded woodpecker

Trapped in a mist net, a woodpecker becomes a research subject in Mississippi. Knowledge of its habits has persuaded some foresters to save its nesting sites: diseased pines. Loss of forest habitat may already have doomed the ivory-bill, officially very close to extinction.

Ivory-billed Woodpecker, Male 1. F. 2 & 3.

PICUS PRINCIPALIS.

With the keen-eyed alertness of birds of prey, a bald eagle glares from her perch in Everglades National Park. Here reproductive success has seemed unaffected by pesticides while nests elsewhere fail. Below, gripping a fish to its body, an eagle rises from a salmon-filled creek in Montana. These birds also eat carrion; at Jackson Hole, Wyoming, one strips flesh from a deer carcass. With nationwide environmental change against it, the bald eagle has dwindled in numbers despite strict protection in federal law.

MARTY STOUFFER, BRUCE COLEMAN INC.

RAMON WINEGAR, BRUCE COLEMAN INC. (ABOVE), AND FREDERICK KENT TRUSLOW (OPPOSITE)

Temblor Range and entered the southern end of the San Joaquin Valley. The change from ridge to valley, from rainy day to sunny day, was dramatically limned by a rainbow.

"So here's this nice valley," Sandy said. "It was in the condor's range. But it offers little food now." Even in the 1930's condors could find the carcasses of cattle and sheep that grazed here. Now alfalfa grows on the rangeland. And, about the time that agriculture came to the valley, the condors in the nearby Sisquoc nesting area began to disappear.

What Sandy believes to be that exiled Sisquoc population was spotted farther north, near San Francisco Bay, in the 1960's. But 10 or 15 birds were mysteriously missing. "It sounds like some kind of sudden mass loss," Sandy told me. He painfully offered a possibility: "Perhaps several of them were seen around a carcass at one time. And perhaps somebody shot them."

In the Tehachapi Mountains, we visited the Tejon Ranch, now a prime feeding ground for the condors of the Sespe sanctuary. Tejon Ranch is one of the largest privately owned tracts of land in the United States—460 square miles. It sustains 14,000 head of cattle, several thousand sheep, two major oil fields, and endless rows of irrigated crops: oranges and cotton, onions and carrots, alfalfa and wheat.

In this empire given so much to the works and financial needs of man I wondered that anyone could spare time to worry about a few oversized vultures. Walter Fieguth, manager of special services and conservation on the ranch, does worry. The ranch cooperates in condor-spotting and the annual count of the birds—a two-day census that involves about 40 observers.

Past a gate kept under lock and key, we drove up a winding, rutted road. Reaching its end, we walked to the rim of a bluff and looked down at the range 5,500 feet below. We had a condor's eye view of the land, crisscrossed by roads and irrigation pipes. We scanned the sky for condor. The wind and rain were up now, and we did not see any. But the sight is not rare. From 1966, when an exhaustive filing system was started, until 1973, condor-watchers have recorded 2,700 sightings.

So the condor still soars. But it is not flourishing. The facts of life about the birds are agonizing: six years to sexual maturity, a single egg precariously laid in a rocky cave, 18 months of care to raise a chick to flight status. With food hard to get, adults don't necessarily starve. But each hour consumed by a distant search for food means another hour of chill for an egg, another hour of hunger for a chick. Such stress seems to be retarding reproduction, naturally torpid even in the best of times.

To ease the food shortage, Sandy, Dean, and John now lug "condor popsicles" to feeding areas. These are road-killed deer accumulated by game wardens, forest patrolmen, and others, and kept frozen until the condor-feeders call for them. The

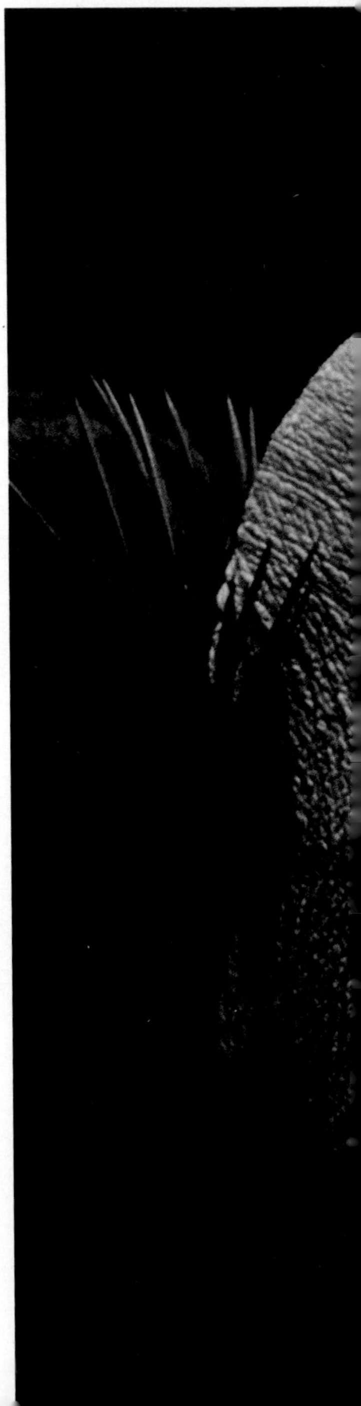

Bald head and sharp bill mark the California condor's adaptation to a diet of carrion—that grows rare as man takes over its range. Wings already of adult nine-foot span, a juvenile soars on updraft currents.

JOHN C. BORNEMAN, NATIONAL AUDUBON SOCIETY (ABOVE), AND J. DAVID SIDDON. (HEAD APPROXIMATELY ACTUAL SIZE.)

carcasses are taken into the rugged terrain of the Sespe sanctuary on "The Thing," a non-motorized contraption that looks like an offspring of a bicycle and a stretcher.

The effort is hilariously heroic. (It inspired some doggerel: "...Dean and Sandy lay in wait/Lenses trained on rotting bait.") But it is also obviously inadequate. Increasingly, the condor needs some kind of management, Sandy says. And it needs continued appreciation for what the bird means to the land and the people of its domain.

"The condor has stopped a $90,000,000 dam project, shifted highways, slowed down oil exploration, and held up, temporarily at least, the prospecting for phosphate in Los Padres National Forest," he recalls. And, as he once wrote, "The California condor may not appeal to everyone as a creation valuable for itself alone, but considered as an ally in keeping California livable, its image changes considerably. Could salvation of our land and our 'life style' depend on a giant vulture?"

The spectacular condor inspires an alliance of impassioned sponsors. Avian supporters rally around such glamor birds as the whooping crane. And when I scan the endangered species list, I see other birds whose death rattles were amplified by friends—and heard by state and federal authorities. The ivory-billed woodpecker, long believed extinct, is still searched for by its never-say-die allies, with the aid of federal biologists. The red-cockaded woodpecker's salvation gave welcome publicity to the U. S. Forest Service and timber companies. The birds' preferred nest sites—old, diseased trees—formerly were cleared away; now foresters retain stands of them.

The large mammals on the list, particularly the wolves and the big cats, are loved by some people and hated by others. These animals generate strong currents of emotion that wash across the list, sweeping some species off and others onto it. The result is a maelstrom of controversy.

Before we plunge into this whirlpool, we should try to sort out the organizations devoted to animals. The groups range from those that do not wish any animal killed to those that actively champion hunting and fishing. The absolute protectionists are deprecated as "doggie woggie groups" by their critics. The latter, who call themselves sportsmen, are often simply labeled "the killers" by protectionists, who, in turn, want hunting and fishing referred to as "blood sports."

Between the extremes are about 40 major organizations with varied interests: the environment, conservation education, game management, non-management, the welfare of animals in general or certain animals in particular. To put it mildly, they don't always agree. Lewis Regenstein, vice president of the protectionist Fund for Animals, used to work for the Central Intelligence Agency. A veteran of many wildlife battles in Washington legislative and executive hearings, he says simply

California condor

In its Sespe sanctuary, near Ventura, an adult lingers at its roost in early morning. A chick some six weeks old backs away from one of its biologist guardians. Condors mature slowly; only 50 or 60 exist.

of his animal work: "This is tougher than the CIA ever was."

No mammal on the endangered list has inspired more debate than the eastern timber wolf, *C. lupus lycaon*. It prowls a narrow, treacherous bio-political terrain. Shot, poisoned, trapped, bountied, and left homeless in state after state for decade after decade, it is making its last stand south of Canada in Minnesota. An unknown number of wolves—estimates range from 500 to 1,000—live in a 12,000-square-mile area that includes Superior National Forest. They are imperiled not by any new encroachments of man, who has snatched away 95 percent of the wolf's historic range, but by his remorseless hostility.

Outright protection of the wolf in Minnesota seemed politically impossible. In 1965, Governor Karl F. Rolvaag vetoed a bill that would renew the state's traditional bounty on wolves. He was defeated for re-election the next year, and many people called his "pro-wolf" stand a factor in his defeat. Four years later, during a poor deer-hunting season, crude signs were posted: "Rolvaag's Wolves Got All Your Deer." Despite evidence to the contrary, many hunters are convinced that wolves seriously deplete deer herds. Linked with farmers and ranchers, the Minnesota hunters form a powerful lobby.

Though the Department of the Interior put the subspecies on the endangered species list, the Minnesota legislature went on record in 1967 saying the wolf was not endangered in that state. The federal-state impasse continued until 1972, when a compromise management plan was developed.

The plan called for raising the wolf's status from unprotected varmint to game animal. Wolves could not be hunted in a "sanctuary" created on federal lands, but hunters could kill 150 to 200 a year, in other areas of the state. Announcement of the plan triggered an explosion of protest. The Fund for Animals, charging that the plan would "virtually destroy the last Eastern Timber Wolf population remaining in the United States," launched a campaign for a more protective version.

In the end, the plan did not go into effect. The Minnesota wolves' fate is still in doubt, but unquestionably they have a host of friends. These include Members of Congress, such as Representative G. William Whitehurst of Virginia. He has not only introduced legislation to protect wolves but also brought one named Jethro to the Capitol to show his colleagues that the big bad wolf is a creature of myths and fairy tales.

I met Jethro, a dignified ambassador of wolfdom and the North American Association for the Preservation of Predatory Animals. Accompanied by his NAAPPA companions, Jethro was continually on tour. He patiently endured hours of gingerly patting by Congressmen, state legislators, three million schoolchildren, and other admirers. He and another ambassador named Clem were taken around the country by John Harris and Anthony Nocera, of NAAPPA, who lives in Brooklyn, New York. On July 28, 1973, back from a tour, Jethro and Clem were in their van outside Tony Nocera's home. During the night someone pried open the door and fed raw chicken to the wolves. The chicken was laced with strychnine. *(Continued on page 170)*

NATHAN BENN

Favorable publicity comes late to the wolf, long an object of fear and hatred, now an animal with active sponsors.

On tour to win friends, Jethro, an amiable Canadian timber wolf, visits a school in Virginia with a spokesman—John Harris, president of the North American Association for the Preservation of Predatory Animals (NAAPPA). Students respectfully keep pace, or peer into a van to watch Jethro and another NAAPPA wolf, Clem.

On the Capitol steps, Jethro demonstrates a wolf's muzzle-to-muzzle greeting with Virginia Congressman G. William Whitehurst, whose wife, Janie, holds the chain collar. Mr. Whitehurst has introduced legislation to protect the wolf. Later in 1973, five months after this trip, a NAAPPA staff man found Jethro and Clem victims of deliberate poisoning in their van. Now Jethro's nephew Rocky, 19 months old, continues NAAPPA's educational programs.

BILL EPPRIDGE (ABOVE, CENTER, BELOW)

MARTIN ROGERS

Barnacle encrusted, a gray whale comes to the surface: a sight to reward "Whale Watchers" putting out from Long Beach, California. Sponsor of these trips—and of whales generally—the American Cetacean Society issues appropriate diplomas. Below, students define a blue whale's size; the car and cans denote heart and aorta. Bulk made the blue a great prize for whalers.

When Tony went out next day to drive the wolves to kennel for a rest, he found Jethro dead and Clem dying.

My daughter Connie had hugged and petted Jethro. When she heard the news, she called me, her voice trembling. "Why?" she asked. I could not answer. But this I know: We have been taught an unreasoned dread of beasts of prey. I like to think that the children who petted Jethro and Clem will not pass on to their children tales of the big bad wolf.

The mountain lion is also stalked by human fears and hates. Theodore Roosevelt maligned it as the "big horse-killing cat . . . the lord of stealthy murder," and people have always found reasons to kill it. The Utah Travel Council urges sportsmen to hunt it there "to collect a seven-to-nine-foot rug or wall piece." The Division of Wildlife Services still looks upon the lion as a subject for "predator control." Experts are unwilling even to estimate how many exist in the United States, and the mountain lion as a species is not on the endangered list. Officials have chosen to list two populations, generally classified as subspecies: the Florida panther and the eastern cougar.

The panther's survival seems to depend on the survival of its lairs—Everglades National Park and national wildlife refuges in Florida. Federal biologists are contemplating a captive breeding and restocking plan for the Florida panther; they do not consider a plan necessary for the mountain lions of the West.

Until very recently, the other cat on the list, the eastern cougar, had little chance for survival because wildlife officials in Canada and the United States were convinced that it did not exist. The story of its return from the dead is largely the story of one stubborn man, Bruce S. Wright, director of the Northeastern Wildlife Station of the University of New Brunswick. "It's not on the way out," he told me. "It's on the way back. The return of the deer is the obvious reason. The cougar is the end of a very short food chain. The soil nutrients and the sun's energy are converted by the vegetation into food for the deer, which in turn converts it into meat and fat for the cougar."

As forests were thinned or hacked away for timber and pulpwood, deer did return. And, Bruce predicted, with them would come "a very few long-tailed tawny cats moving at night from hill to hill." In March 1947, Bruce came upon a male cougar's tracks, which led to tracks of a female and a cub. "Proof of a *breeding* population!" he enthusiastically exclaims a quarter of a century later.

Under ideal conditions—light fresh snow—he and his two companions followed the family's tracks for about a mile. He photographed the tracks, then in July returned and made plaster casts of the male's track in soft ground. Positive identification was made at the U. S. National Museum in Washington.

The proof implied by the cast was not enough; Bruce's inquisitors wanted a body. He could not even provide a photograph. As far as he knows, no eastern cougar has ever been photographed alive. All he could do was accumulate reports of

JAMES A. MATTISON, JR., M.D.

STAN WAYMAN

sightings. By 1960 he had been told of more than 200 reliable sightings in New Brunswick and had reports and track casts from Nova Scotia and the Gaspé Peninsula. His book *The Eastern Panther*, published in 1972, became his ultimate brief in the matter of the missing species. Detailing 304 observations in Canada and 43 in the northeastern United States, he rested his case. In June 1973, the eastern cougar was added to the United States' endangered species list. And Bruce was notified that his country accepts the animal as existing—and endangered.

Fanatical sponsorship of an endangered species for its own sake stands out because that's not the normal way that animals get the attention they may deserve. The people who have given the most support to American wildlife are people who kill animals. In 1973, the Department of Interior distributed to the states $26,500,000 for land acquisition and research "to aid in managing game species." The states were also given $2,100,000 for "hunter safety programs." The federal funds were, in effect, being given back to hunters, for the money came from an 11 percent excise tax on arms and ammunition and a similar 10 percent tax on handgun sales.

Interior also distributed $5,500,000 for sport fish restoration work by the states. This money comes from a 10 percent excise tax on fishing tackle. Sportsmen say they pay their dues, and pointedly ask non-hunters and non-fishermen—usually in the vanguard of protectionist groups—where their dues are.

Some hunters also see the endangered species list as a threat. I have heard their arguments, which are usually off the record. One that wasn't appeared in an October 1973 letter from the secretary of the National Rifle Association to its members: "This hunting season, *you* are the prey. . . . The well-organized and financed gun-control forces have gotten together with the anti-hunting movement. . . . Together, they want to strip us of our rights to own and use guns for self-defense and sport. . . ." I had met this argument before—that non-hunters are anti-gun advocates at heart. This may well be true of some, but the larger, more easily documented truth is that wildlife management philosophies are changing.

National wildlife refuges, many of them operated in the past as game reservoirs, are experimenting with an incentive plan to change management practices. Managers are rewarded for protecting endangered species, for running their refuges as ecosystems, and for attracting "esthetic" users, such as bird watchers. Hunting guides are learning that some people will pay them for showing where to point a camera at an animal—a fact long familiar to guides in Africa, at such famous parks as Serengeti. Several state game departments are beginning to care about non-game animals.

Sea mammals, which once had only the lethal sponsorship of exploiters, now are achieving esthetic status. In San Diego,

With deft forepaws, sea otters feed extensively on shellfish; at left, one finishes an abalone. Scenes like this entertain visitors to Monterey, California, and help recruit allies for "Friends of the Sea Otter."

FRED BRUEMMER (BELOW)

Volcanic ledges of Amchitka Island provide perches for Steller sea lions, now abundant in the Aleutians but greatly reduced in numbers after Bering's explorations in 1741 revealed Alaska's wealth of furs. At right, on Labrador pack ice, a harp seal pup

NATIONAL GEOGRAPHIC PHOTOGRAPHER ROBERT F. SISSON (ABOVE) AND JOSEPH S. RYCHETNIK

shows the trustfulness that brings it crawling to a stranger. A missionary translating the Bible used its Eskimo name kotik to render the term "Lamb of God." Sealers kill pups with clubs for the white fur, worn less than 26 days in infancy; televised scenes of this stirred such revulsion that Canada curbed the hunting of "whitecoats" in 1969. International agreement may prove crucial for this species; a four-nation treaty signed in 1911 rescued the Pribilof fur seals of the Bering Sea. At left, one barks a warning.

boatloads of paying customers head out to view the annual migrations of gray whales. "Whale watching" has become a tourist industry. Sea otters, believed wiped out by 19th-century fur traders, were rediscovered near Monterey, California, in 1938 and eventually made themselves a tourist attraction. "Otters," wrote one of their many fans, "act like they know lovableness is their only weapon."

On the municipal pier at Monterey one morning, I took up watch with Judson E. Vandevere of Stanford University, a renowned authority on sea otters and expert witness for Congress on their behalf. For two hours we saw comic variations on the otter's deft procedure for opening a clam—placing a rock on the chest and pounding the shell open against it—with squabbling sea gulls vying for the scraps.

Then Jud took me out to a windswept bluff on the Monterey peninsula, set up a spotting scope, and aimed it at the dark figures in the glistening blue waters below: sea otters cavorting with their young. We took turns looking. "They're probably rolling to lower their body temperatures," Jud explained as I peered through the scope. "A mother won't roll with a pup on her chest—she takes it in her mouth or holds it in the water. She never has more than one, and it seems to stay with her almost a year." He counted 34 adults, 3 juveniles, 7 pups.

These appealing animals have their own pressure group, the Friends of the Sea Otter, to stave off threats from such commercial interests as abalone fishermen, who blame a shortage of shellfish on the otters' appetite. Other mammals have similar champions, such as the American Cetacean Society, which concentrates on whales, dolphins, and porpoises, and the Fund for Animals, which has dramatized the killing of fur-bearing seals. But the ultimate sponsorship for any endangered species is special legislation. United to save marine mammals, conservationists launched a crusade for a federal law.

The friends of the animals of the sea persuaded Congress to pass the Marine Mammal Protection Act, signed into law on October 21, 1972. With this law came an indefinite moratorium on the killing of marine mammals except by special permission.

The new law explicitly names the species given protection. Responsibility for their protection is split between departments. The law makes sea otters, walruses, polar bears, and manatees the wards of Interior; whales, porpoises, seals, and sea lions become the wards of Commerce.

Would the new law aid wildlife or commerce? An answer came with the appointment of biologist Victor B. Scheffer as chairman of the three-member Marine Mammal Commission that oversees the enforcement of the law. He has written: "If you believe that human life has meaning or purpose or direction or destiny, you will know in your heart that our life is bound all around and together and forever with the lives of the animals who were present at our creation. If we survive, we will care for whales and the other wild creatures, and if we perish through our own cleverness, the end of the wild things will have been an early warning of our folly."

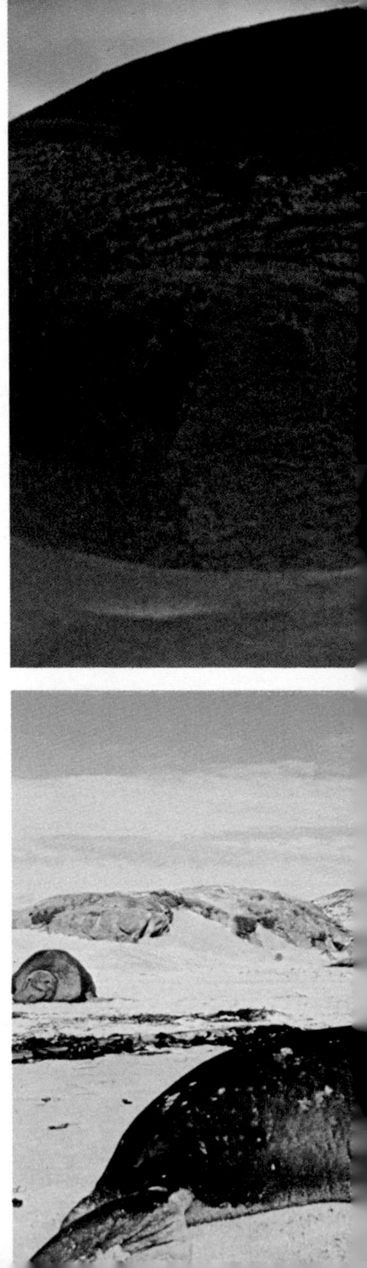

His relaxed proboscis touching his lower lip, a bull elephant seal lifts his head from wind-driven sand on a California island. In a fight for a harem, males inflate the proboscis to snort loud challenges. A "beachmaster," or dominant male, carries the scars of battles won. Timely protection by Mexican laws of 1911 and 1922 saved the last herd of the species; its descendants returned north.

ROBERT B. EVANS

8. Warning to an Endangered Species

On this continent—and this planet—what threatens one may threaten all

Scores of terns swooped down upon us, screeching and pecking our heads in defense of their nests. The air was filled with the cries of outraged birds and the shouts of harassed humans: "Two-four-nine-four. Zero." . . . "Two-three-five-one, marking off with two." . . . "*New nest!* Two-seven-two-two, marking off with two roseates." The six of us stepped gingerly along the rocks and sand in a rough skirmish line, our bombed heads down. We were looking for evidence of something that is usually invisible to anyone but a tern: its nest.

Amid the clamor walked Miss Helen Hays, who carried a clipboard on which she recorded the results of the search. She wore a battered, splattered straw hat, and her shirt and cut-off jeans were redolent of essence-of-tern. Somehow, she managed to look calm and dignified.

The shouted numbers translated into an amazingly complete record of thousands of birds. In nest number 2494, for instance, "zero" meant no new eggs. Nest 2351 had two eggs, and "marking off" meant no need for further checking now. The new nest got its number, 2722, which was written on a tongue depressor and placed near it; it was being marked off with the finding of

Evening mist engulfs New York's Great Gull Island as a volunteer examines a common tern. Unique, detailed records of the tern colony here may indicate the dangers of contaminants in local waters.

BIANCA LAVIES, NATIONAL GEOGRAPHIC STAFF

two roseate tern chicks. Most of the adults we had seen, and suffered under, had been common terns.

Helen Hays and a team of volunteers were hunting for—and worrying about—every egg in every nest on Great Gull Island, a 17-acre dot of land in Long Island Sound. Each day during the May-to-August breeding season she and the crew make a "chick check," a four-hour search for tern nests, eggs, and chicks in the rocks, on the sands, and in the crumbling ramparts of an abandoned fortress. From the Spanish-American War through World War II, the island's Fort Michie had been part of the defense perimeter of the East Coast. If an invader appeared, the fort would flash an early warning.

Through all the wars the alert never came. Now, Helen Hays believes, the island has finally sounded an early warning. The invader is invisible and unknown. But something seemed to happen to terns on Great Gull Island in 1970. And it may be a warning of something that can happen to our children.

Abnormal chicks hatching: some lacking feathers, some with crossed bills, one with underdeveloped legs, one with four legs. The abnormalities suggest those of human "thalidomide babies," who were deformed before birth as a result of medication their mothers took during pregnancy. Scientists call such a deforming chemical agent a teratogen, literally something that produces monstrosities.

The tiny, pathetic monstrosities of Great Gull Island are few. But their number is not as significant as their existence. They probably would not have been found at all if they had hatched anywhere else. For the study of the island's 7,000 terns is so thorough that hardly a chick can pip from an egg without being spotted, banded, and chronicled in years of records that have passed beyond mere human ken and are gorging a computer.

"We're interested in following individuals," Helen says, in modest explanation of the unique study she has been conducting since 1969. It seeks nothing less than embryo-to-death knowledge of every tern hatched on Great Gull Island. The key to that knowledge is an ingenious system of identifying birds with various combinations of color bands.

An observer on the island—or in the Caribbean or South America, at the terns' wintering areas—can correlate a bird's colored bands with records that will tell the exact location of its natal nest, when its mother laid the egg from which it hatched, when it hatched, when it fledged, whether it was the first to hatch, what was the fate of its nestmates. "A bird becomes an individual the minute it has a color combination," Helen says.

The island has been laid out in 25-by-25-meter grids. Many of the rocks, piled around the shore by Army engineers to curb erosion, are numbered. So a discovered nest—a scrape in the open sand or deep in a cleft of the rocks—can be located by a grid number and by a number on the tongue depressor placed near it. The first discovered egg is marked with a "1," the second with a "2." A "3" is rare for roseates. When Dr. Kenneth C. Parkes, one of Helen's chick-checkers, called out, "Two-

JOAN S. BLACK

BIANCA LAVIES, NATIONAL GEOGRAPHIC STAFF

JOAN S. BLACK (THE THREE PICTURES BELOW)

Among splattered rocks on Great Gull, researcher Helen Hays finds a juvenile roseate tern old enough to wear the four coded bands that identify each individual bird. Varied combinations of the bands give keys to life histories kept as computer data. For one such record, naturalist John Hay marks an egg in a newly discovered nest. In a ritual of courtship, a male common tern stands on his mate's back; a female shelters her two-day-old offspring under a wing. Since 1970, reproductive aberrations discovered here have aroused scientific concern: an egg without a shell, a common tern chick with miniature eyes, a roseate chick with deformed legs.

According to Miss Hays and her colleague Dr. Robert W. Risebrough, synthetic chemical compounds—that apparently reach the terns through the fish they feed on—"may, someday soon, result in an increase in the number of human birth defects...."

ARTHUR SINGER

BIANCA LAVIES, NATIONAL GEOGRAPHIC STAFF

three-ooh-two, marked off for roseate with two, has *three,*" Helen shouted back: "Three? Marvelous!"

Colonies of common and roseate terns claimed the uninhabited island until the Army began building the fort in the 1890's. By the time the island had gone through its wars and was retired by the Army, the terns had long since disappeared from it. They also had deserted much of the developing shoreline of Long Island and Connecticut. The American Museum of Natural History acquired the island in 1949 and the Linnaean Society of New York began trying to restore it as a breeding site.

In 1955, about 25 pairs of terns came back. By 1964 the population was stable enough—with about 3,000 breeding pairs—to permit a research group to work in the colony. The museum invited Helen to visit the island that year, and she has been coming back every summer since.

The island remains strictly a habitat for birds. Their watchers live without running water or electricity; a pump or generator might disturb the terns. An ever-changing corps of volunteers camps in ruins that could easily provide the setting for a science-fiction thriller about the end of a civilization.

The setting could not be more appropriate for Helen Hays's worries, which she shares with Dr. Robert W. Risebrough, a chemist at the University of California's Bodega Marine Laboratory and a veteran of pollution-in-wildlife mysteries. The two scientists warn that the terns of the abandoned fortress are now inadvertently exposing "a newer, more insidious enemy to man: contaminants. Presently disturbing the reproduction of the terns, these contaminants may, someday soon, result in an increase in the number of human birth defects as they work up through the food chain to the ultimate consumer."

What are these contaminants? The list of possibilities includes DDT, related pesticides such as aldrin and dieldrin; mercury and other heavy metals—and a relatively new suspect: the polychlorinated biphenyls, known in chemical shorthand as PCBs. Versatile compounds useful in everyday products like wax, varnish, and adhesives, PCBs help retard fire and moisture damage, resist heat, increase the effectiveness of pesticides. They have been important ingredients in many plastics.

"The pathways by which PCBs escape into the ecosystem are poorly known," reported scientists David B. Peakall and Jeffrey L. Lincer from Cornell University in 1970. "The evidence suggests that it is important to find the leak and stop it." In the previous 40 years, an estimated one *billion* pounds of PCBs had come to rest in dumps or escaped into the ecosystems of the world. Of that billion, about 600 million pounds may be affecting the environment in the United States. In 1971 the producer, Monsanto Company, voluntarily agreed with the Federal Government to limit PCBs to essential "closed-system"

STEPHEN S. BAINBRIDGE

Abandoned as a chick, a young Bermuda petrel flexes wings ready for flight; Bermuda conservation official David Wingate, who reared it, takes a final record of weight. The adults spend almost all their lives at sea; a specialist calls them ideal monitors for oceanic pollutants.

uses, hoping to stop the leak here. But it seems clear that PCBs are already as widely diffused throughout the world as DDT.

Helen's colleague in the early-warning report, Dr. Risebrough, became especially concerned about polluted environments when he investigated the status of the brown pelican at traditional nesting sites on the West Coast. The big-pouched bird began vanishing from the Gulf Coast in the 1950's. Along the Texas and Louisiana coast the pelican population dropped from about 50,000 to fewer than 100 in the mid-1960's. Researchers were worriedly checking chicks and eggs.

They were looking for what ornithologists began finding with appalling frequency in the late 1960's: mysteriously thin-shelled eggs. Biological detective work proved that DDE (a derivative of DDT) disrupts shell formation in certain birds. This was verified by experiments and comparisons of fragile new eggs with eggs of the pre-DDT era, in museum collections.

The pelican's decline was not confirmed on the West Coast until March 1969, when Dr. Risebrough inspected a rookery on west Anacapa Island, at the southern end of the Santa Barbara Channel. Of 298 nests in the colony, only 12 held intact eggs. In June, he returned. He found a single hatchling.

The Fish and Wildlife Service estimates that more than 40 species of birds suffer from eggshell thinning. Most of these birds live at the top of food chains; often, the longer the chain, the greater the potential dose of poison for the top consumer.

By a process known as biological magnification, poisons that begin at the bottom of a chain build up on their way to the top. At Clear Lake, California, for example, an insecticide was applied at what seemed just enough strength to wipe out gnat larvae. But the plankton in the lake concentrated this poison at a level 265 times higher than that in the water. Small fishes that ate the plankton had a poison load 500 times more concentrated than the original dosage. The top-of-the-chain consumers were western grebes, long-necked birds that feed by diving for fish. Concentrations in the fat of two grebes were "magnified" 80,000 times. Some birds were killed outright; a breeding colony of two thousand disappeared.

On Gardiners Island, near Great Gull, what may have been the world's largest breeding colony of ospreys plummeted from 300 pairs in 1945 to 31 in 1973. Roger Tory Peterson, lamenting the loss of ospreys near his home in Old Lyme, Connecticut, reported a drop from 150 active nests in 1954 to 10 in 1968. By 1973, only two pairs nested on the Connecticut River.

A nationwide cutback on DDT, ordered in 1972, has not erased the persistent pesticides from the land. Dieldrin, applied as such or derived from aldrin, lingers in soil, is transported by winds and waters, finds its way into the tissues of animals from earthworms to whales and human beings. (Pelicans, however, show feeble signs of possible recovery: Texas in 1973 celebrated the discovery of 11 fledglings.)

The twilight of DDT comes as a new monogram—PCB—looms over the environment. *(Continued on page 186)*

Menace of mercury: Derived from household and industrial wastes, potentially fatal methyl mercury compounds course through the food chain of a freshwater ecosystem, lodging finally in man. Dumped as garbage, an array of everyday products—paints, cosmetics and pharmaceuticals, batteries, disinfectants, soaps—contains mercury; mercury-bearing effluents from factories settle into the water. Organisms absorb it directly from the water. It progresses through the food chain as one creature eats another, beginning with plankton—microscopic plant and animal life, here greatly enlarged—and passing to larger and larger fish. Mercury re-

PAINTING BY ARTHUR LIDOV

mains within the organs of the body; research indicates that its levels increase as it moves from animal to animal. It may contaminate these typical freshwater game fish (from left, clockwise, but not to scale): yellow perch, bluegill, carp, smallmouth bass, walleye (swimming downward), channel catfish, northern pike, lake sturgeon. A fisherman, end point of both food chain and mercury accumulation, displays a pair of catfish. Although no confirmed human deaths have resulted from mercury-poisoned fish in the United States, at least 52 Japanese have died—and 80 now suffer disabilities—from eating seafood tainted with mercury.

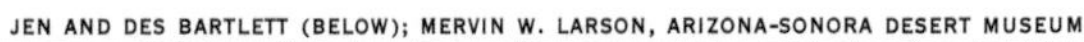
JEN AND DES BARTLETT (BELOW); MERVIN W. LARSON, ARIZONA-SONORA DESERT MUSEUM

Consistently hungry but not greedy in manner, an osprey chick plucks food from its mother's beak; parents built the clifftop nest on a Mexican island in the Gulf of California. In swift dives, ospreys snatch fish from the water with their talons and return to the nest; the male eats the head, his mate feeds the body to the young and gets the tail or other remains herself. On another islet, an osprey in juvenal plumage spreads its wings. The ultimate consumers in their food chain, ospreys in some areas show the effect of such accumulated pesticides as DDT: reproductive failure, marked by eggs so thin-shelled that they collapse. Osprey researcher Paul Spitzer of Connecticut says of this threat: "Ospreys are more than just birds to be enjoyed. They are an alarm system of things gone haywire in the river, the estuary, and the sound. They are sensitive indicators of the environment."

Concentrations of PCBs were higher than DDE in the terns of Great Gull Island; the fishes they ate carried PCB residues.

To find out what this means to us and our children, we have to follow scientists from Great Gull to the West Coast, from laboratories in the United States to hospitals in Japan, and into the structure of chromosomes and cells that we share with all living things. Like the radiation from atomic explosions, certain chemicals can damage chromosomes. The changed chromosomes, through the genes they carry, may pass changes—mutations—from one generation to the next; the carrier of mutated chromosomes may show no effects, but the carrier's offspring may show, or bear undetected, a burden of mutations.

No matter what the animal, the mechanisms of mutation seem to be the same. So, in higher organisms, is the marvel of development. Birds, fishes, reptiles, amphibians, mammals—all start the same way: as a fertilized egg, which divides into two cells and continues to grow as the cells multiply. And at this point, teratogens can induce damaging changes whether the embryo is developing in an egg or in a womb.

Because of this universality of life, studies of the effects of PCBs on one species may reflect what such chemicals can do to other species. Or the results may remain inconclusive, because different species show varying reactions. Yet PCBs that invade life systems seem capable of inflicting damage on embryos and on chromosomes.

In one experiment, six pairs of ring doves, kept in a Cornell laboratory, produced 24 eggs; 22 hatched and all 22 hatchlings fledged. Dr. Peakall and Dr. Lincer fed the parent doves PCB-laced food for only three months and then let them incubate a new clutch. Now the reproduction record was 24 eggs laid, 24 chicks hatched and fledged. Six months later, birds of this generation were bred; of 20 eggs laid, only 4 hatched. One hatchling lived one day; one, deformed, was sacrificed for study at three weeks. Two fledged.

PCB levels in the muscle tissue of the abnormal tern chicks from Great Gull, the Cornell scientists noted, ran even higher than the levels in the experiment's adult ring doves. Examination of the ring-dove embryos showed "a possible clastogenic (chromosome breaking) action of PCBs." Such "chromosomal aberrations," the Cornell report said, could be the cause of the abnormalities in the terns of Great Gull Island.

Sea mammals carry their young as do human mothers, in the womb. To threaten the unborn, chemicals must somehow pass the placental barrier that guards it within the mother or disrupt the normal term of pregnancy by affecting her hormone balance. These threats are apparent in harbor seals of Washington's Puget Sound, in sea lions of the Channel Islands strung along the California coast from Santa Barbara to San Diego.

Pups are being stillborn, or born prematurely, soon to die. On San Nicolas, one of the Channel Islands, a zoologist counted 135 aborted sea lion pups in 1969; in 1970 he counted 442, and the cry of "early warning" went up. To the northwest on San Miguel, a sanctuary for foxes *(Continued on page 193)*

G. RONALD AUSTING, NATIONAL AUDUBON SOCIETY (ABOVE); GUS WOLFE (OPPOSITE); ANTHONY MERCIECA, NATIONAL AUDUBON SOCIETY

Home after hunting to feed his young, age about 18 days, a male prairie falcon stares watchfully from a rock-ledge nest. Age 38 days, a nestling squawks noisily as it settles on a scale for a Montana study of growth rate in prairie falcon chicks. Trained to hunt for sport, a peregrine falcon stands over his prey—a mourning dove. Pesticides jeopardize all the American falcons, with the peregrines' decline the best-known.

snowy owl

screech owl

great horned owl

Gliding through the Ohio night on wings that spread four feet, a great horned owl coasts toward its nest after hunting small mammals. Three-week-old owlets will share a rat killed by a parent. In Georgia's Okefenokee Swamp, a screech owl blinks an eye—large and adapted for night vision; a snowy owl rests in a Connecticut salt marsh. At the top of terrestrial and relatively short food chains, owls may ingest less pesticide than birds that prey on fish.

G. RONALD AUSTING, BRUCE COLEMAN INC. (ABOVE); BELOW, FROM LEFT: STEPHEN COLLINS, PHOTO RESEARCHERS, INC; JEN AND DES BARTLETT; RON KLATASKE

CARLETON RAY, PHOTO RESEARCHERS, INC. (OVERLEAF); DAVID HISER (ABOVE AND LOWER LEFT)

Overleaf: Lord of a frigid realm, a polar bear surveys the Arctic wastes for other bears—or for man—from an ice floe in Canada.

Immobilized by a drug, a male polar bear sprawls in the snow as biologist Charles Jonkel of the Canadian Wildlife Service clips an ID tag to its ear—and returns to his helicopter. A cub 14 weeks old clambers onto its tranquilized mother. Such studies reveal that contaminants have entered food chains even in the remote Arctic. At a bear's—or man's—kill, the ivory gull eats scraps and bloody snow; the quest for oil threatens its only known North American breeding ground, Seymour Island.

S. D. MACDONALD, NATIONAL MUSEUM OF NATURAL SCIENCES, OTTAWA (CENTER); THOMAS B. ALLEN, NATIONAL GEOGRAPHIC STAFF

as well as sea lions, colleagues found 242 premature pups on one day in 1970 and 348 in 1971. They examined four females with healthy full-term pups, six that had just aborted their young. The only significant difference: The six aborting females had 8.5 times as much DDT and PCBs in their blubber as the others, in addition to high levels of mercury.

Terrell C. Newby began studying harbor seals in Puget Sound in 1964 and continued his work as a doctoral candidate at the University of Washington. In a population of about 200 on Gertrude Island, he has seen the birth-defect rate soar from near-zero to 10.5 percent in 1970 . . . to 13.3 percent in 1971 . . . to 37 percent in 1972. The pups' tails were twisted or missing, their skeletons deformed. Some had incomplete backbones or intestines. "PCBs," he says, "look very much like the problem." They have been found in newborn pups—and could have reached them only by crossing the placental barrier.

From Japan there came a link to man. Terry Newby points to a new disease that Japanese call *yusho*, poisoning by PCBs. Mothers who ate rice oil accidentally contaminated with PCB compounds gave birth to smaller-than-normal babies with blotched skin, skull defects, or liver, mouth, and blood disorders. PCBs were detected in placentas and in a stillborn baby. "It is clear," Japanese researchers reported, that PCBs were "transferred [to the fetuses] from their mother via placenta."

The realization that threatened wildlife can signal early warnings to man has inspired a search for animal monitors, modern versions of the coal-mine canaries whose deaths alerted miners to lethal fumes.

One monitor of pollution—past, present, and future—is the peregrine falcon. Two subspecies, the American peregrine and the Arctic, are on the endangered species list. The American, virtually extinct in the eastern half of the continent, rapidly disappeared in the 1950's. An early victim of eggshell thinning, it topped a long food chain.

In 1970, a two-nation team of investigators checked 237 known falcon eyries in Canada and Alaska. The empty nests they found reveal the spread of poisons northward. In southern Canada, only four pairs were found in 82 recorded eyries. In the Arctic tundra, 53 eyries yielded 31 pairs. "Most of us like to think," the joint report concluded, "there are vast stretches of Canada and Alaska that are still pristine, undisturbed wilderness, providing a haven in which Peregrines can thrive indefinitely." But "the world is a chemically polluted environment. . . . And so the Peregrine continues to disappear at a rate that could bring it to extinction in North America in this decade."

There seems to be no escape, not even in the Arctic, not even for the topmost animal of the food chain at the top of the world, the polar bear. Within its awesome body it carries the chemicals known in so many others so far away. PCBs have been found in polar bears.

But this is no canary warning us of the first whiff of death. This is no winsome seal dying as a harbinger of the death that may cross barriers to us. *(Continued on page 198)*

DAVID HISER (ABOVE AND BELOW)

DAVID HISER

From a 50-foot crag of ice, tip of a berg trapped in Jones Sound, three of Dr. Jonkel's field party—including Lazarus Kayak (left)—scan the pack ice for bears to tag. Nearby, musk oxen gallop across a snowy plain typical of their primeval range in Alaska, Canada, and Greenland. Some 17,000 musk oxen thrive under government protection. But today Arctic animals—musk ox and bear among them—face habitat disruption as developers of oil and natural gas resources invade their domains. In recent years the Inuit—"the People," known as Eskimo—once totally linked to the polar animals, have altered a mode of life centuries old; they no longer depend totally on nomadic hunting. Already the changes that have upset the wildlife balance elsewhere in North America begin to overtake the complex web of life in the harsh but delicately poised environment of the Arctic—a land that the author calls the "last frontier of the continent's wildlife."

American alligator

Texas blind salamander

Colorado squawfish

unarmored threespine stickleback

Reptiles, fish, birds, mammals—an assembly of North American wildlife confronts the danger of extinction. In this grouping of representative species, those spectral animals in the dark band have already vanished; others—some critically endangered—cling to existence in small numbers. Often, like the whooping crane, they survive in protected enclaves. Timely management of some animals—the bison and the wood duck among others—has stabilized a once-threatened species. A 1972 federal law protects such marine mammals as otters, whales, and polar bears.

PAINTING BY JAY H. MATTERNES

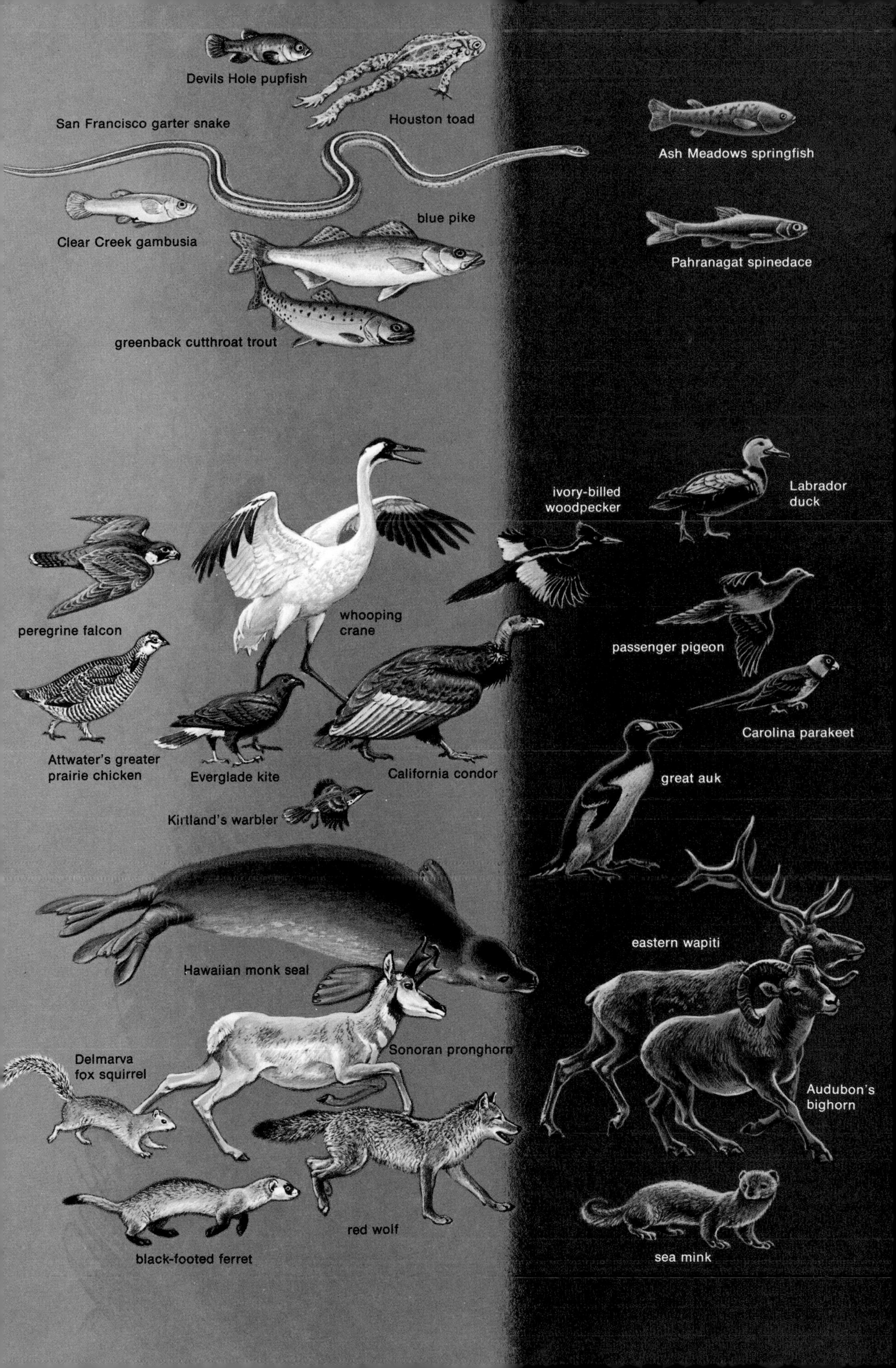
Devils Hole pupfish
Houston toad
San Francisco garter snake
Ash Meadows springfish
Clear Creek gambusia
blue pike
Pahranagat spinedace
greenback cutthroat trout
ivory-billed
woodpecker
Labrador
duck
peregrine falcon
whooping
crane
passenger pigeon
Attwater's greater
prairie chicken
Everglade kite
California condor
Carolina parakeet
great auk
Kirtland's warbler
eastern wapiti
Hawaiian monk seal
Sonoran pronghorn
Delmarva
fox squirrel
Audubon's
bighorn
red wolf
black-footed ferret
sea mink

DICK ROBINSON, PHOTO RESEARCHERS, INC.

This is the ultimate animal, towering over the continent. And it lives, prevailing not as a symbol of the imminent death of wild life but as a symbol of the harshness of existence, the endurance of all life. I went to the Arctic to see a species in peril. I saw instead an animal that still prevails. Man, invading this last frontier for its shrouded riches, has not yet struck down his only challenger.

I began learning about the confrontation of men and bears soon after I arrived at Resolute, a Canadian outpost more than 500 miles within the Arctic Circle. Here men have watched the birth of the continent's weather and here men now launch deeper penetrations of the Arctic in search of oil and natural gas.

One of the few men in Resolute for other reasons was Dr. Charles Jonkel, a research biologist for the Canadian Wildlife Service. Waiting for the skies to clear for a helicopter trip to tag bears and search for den sites, he talked with me about men and bears. "When man decides to conquer an area," he said, "the first animals he destroys are those that compete most with him for space—the large carnivores, the wolf, the bear. Now that the Arctic, the High Arctic, is being invaded by man, he must compete for space with the largest carnivore here. I think that the polar bear could teach man a lesson. They can live together here. And if the white man can learn to live with bears, maybe he can learn to live with other men—white men and black men or yellow men."

In much of their circumpolar range, polar bears have belatedly gained protection. (Reported kills in Canada alone rose from 148 in 1940 to 509 in 1960). In 1965 the "polar bear nations"—Canada, the United States, the Soviet Union, Norway, and Denmark—decided to protect mothers and young, and to foster research. In late 1973 they worked out a formal conservation agreement providing for stronger protection, mutual consultation on bear management, and coordinated research programs.

Of the world's polar bears, 20,000 or so, the majority live in Canada, which has an annual hunting quota of more than 500. Although specialists argue the matter, Dr. Jonkel believes the bear is now safe in most of its range. He told me, however, that one "extensive oil spill could change this instantly."

He stabbed a green-stained finger at a map. "There's oil well number one. And number two and number three." He traced the probable path of a pipeline that would carry natural gas. He hopes that information he is gathering on denning areas will divert the pipeline. "We're trying to teach people that the Arctic isn't just a sheet of ice," he says. "It has abundant wildlife. There's only one Serengeti in the world. And there's only one place like this in the world. If anyone can do anything to protect key portions of this sort of habitat, Canada's got to do it."

Two weathered-in days later, we piled into the helicopter:

Lunging backward, a grizzly bear grasps a large salmon in its strong jaws after killing it in a Wyoming stream. Such scenes evoke the heritage of North American wildlife—a natural heritage of freedom.

engineer-pilot Jim Gray, Frank Brazeau, an ex-lumberjack who had joined the Canadian Wildlife Service; Chas Jonkel (his nickname rhymes with jazz), and I, bundled more than the others against my first exposure to 30-below-zero cold. Forty miles out of Resolute, we spotted our first tracks, a straight line of immense paw marks startlingly visible against the rippling white infinity. In minutes we had swooped down on the bear and landed nearby to estimate its weight for drugging.

Chas swiftly loaded a rifle with a tranquilizing dart and we took off. As we trailed the tiring bear, Chas opened a door, leaned out, and fired. The bear looked up at us and loped away. We landed again, waited a few minutes for the drug to take effect, flew to where it lay, and landed. Jim stayed aboard. I followed Chas and Frank. The bear was conscious, growling feebly. As they worked, the bear had a convulsion; so did I until I was told that the bear would not get up. It was a male which weighed 500 pounds, an estimate based on the girth of its chest as measured by a tape marked off in pounds. I wondered about the tape's accuracy. Later, Chas measured me: three pounds off. "Works on just about any large mammal," he said.

Back in the helicopter, and soon two more bears—a mother and cub. Chas darted only the mother, and I soon saw why. The cub, about three and a half months old, would not leave his fallen mother. He did not run away, Chas explained, because he responds to his mother's cues. Tranquilized, she showed no fear, and so neither did her cub.

Chas and Frank attached blue tags to the mother's ears. Chas stained his fingers a deeper green as he imprinted with tattoo-needled pincers a four-digit number inside her lip. He then pulled a tiny premolar tooth; its growth rings would reveal her age. Frank straddled the struggling cub to hold it down for tags and tattooing. The cub kept all his teeth. Chas knew that it must have been born around Christmastime.

Polar bears have an adaptation that biologists call delayed implantation. The bears mate in May, but the fertilized egg does not start developing until around September. About two months later, the mother digs a den, usually in a deep snowdrift, and stays there until her young—twins, as a rule—weigh about 20 pounds. They are born blind, weighing about a pound and a half. Mother and young leave the den in March or April.

The mother often sculpts a two-room den with an entranceway located so that cold air does not blow into the inner rooms. Sometimes she carefully claws the ceiling, creating a kind of translucent sunroof that produces a greenhouse effect. The earliest igloos may have been modeled after polar-bear dens.

Toward the end of the day, Chas spotted a hole in a snowbank high on a steep hill. We circled it, landed nearby. Chas, Frank, and I got out of the helicopter. As usual, Jim stayed aboard, keeping the engine running. I watched Chas approach the hole in the snowbank and poke his head into it. Frank, who was new to this work, had gone uphill, beyond Chas. Now he walked toward me and I walked toward him. Chas was backing out of the hole. I was just about 20 feet from it when . . .

Droplets of water tossed high by cavorting youngsters catch afternoon sunlight in the Minarets Wilderness of California. A six-year-old camper on a pack trip ladles boiling soup into his Sierra Club cup. In such protected areas as a national wilderness, animals still roam free; man comes to these lands only as visitor or guardian.

By the end of this century, according to some predictions, these preserves may hold the only remnant enclaves of the rich fauna that settlers found on this continent. The Reverend Dr. William G. Pollard, scientist and priest, foresees that in those days man will "sense the pathos of a vanished world. . . ."

All in an instant Frank drops from sight and yells, "She's got me! She's got me!" The head and upper body of a polar bear appear. Frank bobs up flailing his hands at the bear's face. The bear swivels her head and sees Chas. He starts running toward Frank. Frank scrambles out of the caved-in snow and stumbles away. Jim, roaring the engine, skims his chopper toward the bear, which ducks out of sight. The bear's head reappears. Jim lands the helicopter near Frank. We get him aboard and, as we lift off, Jim radios the base at Resolute for medical aid.

"He's OK," the nurse at the Eskimo settlement near Resolute said. "Two stitches." While the nurse worked on Frank's wounds, Dr. Jonkel—I couldn't think of him as Chas just then—looked them over. The deep one, gleaming on Frank's leg like a red nailhead, was made, Dr. Jonkel said, by one of the bear's four canines. She had penetrated, not quite as deeply, with two other canines. "But," Dr. Jonkel clinically noted, "she wasn't able to cross them and take a bite of meat." He explained later that a den would lie uphill from its entrance hole; he had missed shouting a warning only by seconds.

Back in Resolute, Frank told us what had happened. He had not fallen into the den. He had been pulled in. "She broke through the snow and grabbed me. I saw yellow fur—and her eyes, eye to eye. I remember leaning back and hitting her. And then running." The bear had not chased him, Chas speculated, because she was defending her den and young.

"I guess," Frank said, "this is the only time a man fought with a bear and they both lived through it." He was obviously glad the bear was alive.

And I know that he and the other admirers of the Arctic are glad about the life they see. There, where life is so precious, where survival becomes as much a part of a man as it is a part of a polar bear or a caribou or musk ox, there is where extinction is the bitterest word of all.

If ever an early warning is sounded there, on that last frontier of the continent's wildlife, it will be the final warning. Nothing will be left for them or for us. Below the frontier, the tomorrows that began at Charles Towne seem to be running out. A forest full of newfound life became a zoo of remembered life.

"With the best that any of us can do," says a man of God and science, "by the end of this century the only wild animals left on the earth will be found in zoos or scattered national parks." In his vision of that day, the Reverend Dr. William G. Pollard, priest and physicist, sees a planet encrusted with man's cities. "Occasionally he will visit a zoo or a wildlife preserve and sense the pathos of a vanished world...."

Eagle and whale, seal and sea otter, polar bear and tern—all are chained to the food and water of a world grown hostile to life. They cannot prophesy. But, flickering out, they can enlighten. We, too, are chained to their vanishing world.

Near summer's end, a backpacker hikes across a hillside of Maroon Bells-Snowmass Wilderness in Colorado—a land that, for the present at least, still harbors a diversity of mankind's companion creatures.

Index

Boldface indicates illustrations; *italic* refers to picture captions.
† A dagger designates extinct species or subspecies.

Library of Congress CIP Data

Allen, Thomas B.
Vanishing wildlife of North America.
"Prepared by the Special Publications Division of the National Geographic Society."
1. Rare animals—North America. 2. Wildlife conservation—North America. 3. Wildlife management—North America.
I. National Geographic Society, Washington, D. C. Special Publications Division.
II. Title.
QL88.A44 591'.04'20973 73-833

ISBN 0-87044-129-9

Additional Reading

The reader may wish to refer to the Society's books *The Marvels of Animal Behavior, Wild Animals of North America, Song and Garden Birds of North America,* and *Water, Prey, and Game Birds of North America,* to check the *National Geographic Index* for related materials, and to consult the following references:

Roger W. Barbour and Wayne H. Davis, *Bats of America;* Kai Curry-Lindahl, *Let Them Live;* Albert M. Day, *North American Waterfowl;* James C. Greenway, Jr., *Extinct and Vanishing Birds of the World;* C. A. W. Guggisberg, *Man and Wildlife;* Oliver H. Hewitt, ed., *The Wild Turkey and Its Management;* Joseph P. Linduska, ed., *Waterfowl Tomorrow;* Faith McNulty, *Must They Die?;* P. S. Martin and H. E. Wright, Jr., eds., *Pleistocene Extinctions;* Peter Matthiessen, *Wildlife in America;* Ernst Mayr, *Animal Species and Evolution;* L. David Mech, *The Wolf;* Morton W. Miller and George C. Berg, eds., *Chemical Fallout;* Wilfred T. Neill, *The Last of the Ruling Reptiles;* Roger Tory Peterson, ed., the Peterson Field Guide Series; Frank Gilbert Roe, *The North American Buffalo;* Robert L. Rudd, *Pesticides and the Living Landscape;* Russell J. Rutter and Douglas H. Pimlott, *The World of the Wolf;* Glen Sherwood, "If It's Big and Flies . . . Shoot It!" in *Audubon,* November 1971; James B. Trefethen, *Crusade for Wildlife;* U. S. Department of the Interior, Bureau of Sport Fisheries and Wildlife, *Threatened Wildlife of the United States;* Bruce S. Wright, *The Eastern Panther.*

Acknowledgments

The Special Publications Division is grateful to the individuals, organizations, and agencies of government named or quoted in this book, and to those cited here, for their generous cooperation and assistance during the preparation of this book: consultants at the National Aquarium, the National Arboretum, and the Smithsonian Institution; public officials at national, state or provincial, and local levels; Defenders of Wildlife; the Wildlife Management Institute; Lt. Ken Alvarez, Dr. Roger W. Barbour, Dr. Gerald A. Cole, Dr. Richard W. Coles, Sr. Salvador Contreras Balderas, Dr. Frank M. D'Itri, Mr. William G. Gilmartin, Dr. F. Wayne King, Dr. Carl B. Koford, Mr. Sterling E. Lanier, Dr. Ernst Mayr, Dr. L. David Mech, Mr. Ralph W. Schreiber, Dr. George Gaylord Simpson, Dr. Charles F. Wurster.

Composition for this book by National Geographic's Phototypographic Division, Carl M. Shrader, Chief; Lawrence F. Ludwig, Assistant Chief. Printed and bound by Fawcett Printing Corp., Rockville, Md. Color separation by Colorgraphics, Inc., Beltsville, Md.; Graphic Color Plate, Inc., Stamford, Conn.; Progressive Color Corp., Rockville, Md.; J. Wm. Reed Co., Alexandria, Va.; and Tyler Graphics, Arlington, Va.

NATIONAL GEOGRAPHIC SOCIETY

WASHINGTON, D. C.

Organized "for the increase and diffusion of geographic knowledge"

GILBERT HOVEY GROSVENOR
Editor, 1899-1954; President, 1920-1954
Chairman of the Board, 1954-1966

THE NATIONAL GEOGRAPHIC SOCIETY is chartered in Washington, D. C., in accordance with the laws of the United States, as a nonprofit scientific and educational organization for increasing and diffusing geographic knowledge and promoting research and exploration. Since 1890 the Society has supported 972 explorations and research projects, adding immeasurably to man's knowledge of earth, sea, and sky. It diffuses this knowledge through its monthly journal, NATIONAL GEOGRAPHIC; more than 30 million maps distributed each year; its books, globes, atlases, and filmstrips; 30 School Bulletins a year in color; information services to press, radio, and television; technical reports; exhibits from around the world in Explorers Hall; and a nationwide series of programs on television.

NATIONAL GEOGRAPHIC MAGAZINE